Miners sluicing for gold.
(*Photo courtesy of Security Pacific National Bank*)

WHERE TO FIND GOLD IN SOUTHERN CALIFORNIA

James Klein

WARD RITCHIE PRESS • PASADENA

FOR MY WIFE WHO,
AFTER TWENTY YEARS
AND SEVEN CHILDREN,
IS STILL THE GREATEST TREASURE
I HAVE EVER FOUND

Maps by Bob Tinnon and Gayle Cottingham
Cover by Mike Dooley.
Endpapers: Unusual old sluice located in
San Francisquito Canyon.
(*Photo courtesy of Security Pacific National Bank*)

The material in this book is reviewed
and updated at each printing

Library of Congress Catalog Card Number 74-28535
ISBN 0378-03111-2
Lithographed in the United States of America

CONTENTS

Mining in the San Gabriel Canyon in the 1890s.
(Photo courtesy of Title Insurance and Trust Company)

INTRODUCTION

Ask any old prospector where the best place to search for gold is and, more than likely, he will tell you, "Gold is where you find it." After you have been prospecting for a while, you will find that no truer words were ever spoken. You can find gold in many places in Southern California. This book will show you where and how, in addition to giving you leads on finding a lost treasure as well.

Almost everyone believes that gold was first discovered in California at Sutter's Mill on the American River by James Marshall in 1848. This is, unfortunately, not true. Nor is it true, as others claim, that the original find was made by Francisco Lopez in 1842 in Placerita Canyon, near Newhall in Los Angeles County. The truth is that by the seventeenth century, when the Spanish arrived in California, the Indians were already using gold for various decorative purposes. When the early missionaries discovered this, they had the Indians lead them to the source of their gold, and then forced the Indians to mine it for them. Many of the lost-mine tales of the West stem directly from Indian revolts against their enforced slave labor. The remains of these early mines still can be found in the deserts and mountains of Southern California.

This book will pinpoint and describe the locations of the gold-bearing areas near the Los Angeles basin and will relate some of the treasure tales associated with each area. None of the areas described is more than a one-day trip for any Southern Californian. Once you get to the gold, you will want to know what it looks like, how to get at it, and what to do with it when you have it. The second section of the book, "How to Find Gold," will answer these practical questions.

I first found gold in Southern California while working as a salesman. Every day, after making my calls, I would head for the nearest gold-bearing area for a couple of hours of prospecting. Many's the time I had to use my headlights to check my sluice box or pan to see what I had found. Sometimes the pickings were good. Other times, I'd just chalk it up to experience.

After several years of mining, I found a lost gold mine. A professional geologist and a licensed state of California assay laboratory have estimated that my mine contains several million dollars in gold and silver. How I found it might interest you.

Many treasure tales are associated with missions such as San Buenaventura.
(Photo courtesy of Security Pacific National Bank)

I located my mine by a combination of research, field work, and some old-fashioned luck. For many years I made it a practice to read anything I could find about gold, silver, lost treasure, and mining in California. I had gone to local libraries and studied the history of known treasure locations, and had talked to local old-timers. The key clues to the discovery of my mine came from an eighty-year resident of the little town closest to the mine. He saved me days of wasted effort by recalling stories his father had told about the mine's location. I had been searching an area nearly a mile from the mine's actual site and I might never have found it without the information learned from this man's conversation. Now, let me tell you the *whole* story of how I found my lost treasure.

On a summer Sunday afternoon outing with my wife and children, I came across an old church sitting off a back road south of Los Angeles. We always have been interested in the history of California and make it a habit to visit any of the early missions or museums we discover in our travels. The church had been a sub-mission to one of the larger missions. It was built to serve the Indians in one of the main church's outlying areas. While visiting, the priest in residence told me an intriguing treasure story connected with the church. Years ago, it seems, the local Indians had hidden two bags of gold ornaments from the Spanish in mountain caves behind the church. Several members of the tribe were tortured and killed for refusing to reveal the location of the treasure. After that, the Indians believed that the ghosts of the dead Indians guarded the treasure, and they were afraid to search for it. I was interested in searching for treasure, of course, but I immediately wondered about the gold source the Indians had used to make the ornaments.

The next day I went to the State Department of Mines and Geology, where I purchased a geological map of the area. These maps are very helpful to the prospector and treasure hunter because they show all known mineral deposits, mines, and prospects. They also show regions that geologists feel are potential areas of mineral deposits still to be located. Looking at the map, I could see that the old church was located near known placer deposits and also an area of potential future gold deposits.

Next, I studied some old reports made by various state mineralogists over the years. After several hours of fruitless searching, I had almost decided to give up when I found it. In a report dated 1893, the state mineralogist had written that there was a little-known, ancient, gold-bearing river channel in the county which had been worked for years by Indians and Mexicans. The gravels, he said, were from an ancient river that must have flowed from the east or the north, over the gold belts, on its way to the sea. The gold was thought to be spread all through the gravels. The deposit was in a ridge 2,000 feet higher than the countryside, but the lack of water made any large-scale production impossible.

After finding this report, I consulted other books about lost treasure tales of the area. I found several slight references to a lost Spanish gold mine near my region. The mineralogist's report had given a general location of the placer field using a farm, a ranch, and a mountain peak as references. The only point I could be sure of was the mountain, since the farm, ranch, and their owners no longer existed. I made many trips to the area with no luck at all.

One day, after hacking my way through what seemed like miles of brush, I quit searching and went into town for some gas and a cold drink. At the general store there, the old fellow I mentioned earlier gave me the clue that led to my discovery. He told me that from the old road to the next town you could still see the remains of an old trail that led to the diggings. Only from a certain point and only at twilight when the light was just right, could the trail be seen. I rode up and down the road several times before I was able to see the trail, and even then it was just a faint tracing on the side of the mountain. All signs of the trail for a couple of thousand yards from the old road to the mountain had disappeared.

Using some prominent objects as guideposts, however, I made my way to the side of the mountain. The trail was almost impossible to pick up when you were on top of it; bushes and even trees now grow right in the middle of it.

After several hours of fruitless searching on ridges around a beautiful valley, I had almost decided to give up, when I found it. I'd found the location of the old workings. I took samples of the gravel deposits to be assayed. As I said, the assays turned out rich. I staked and filed my claim, and I am now working with a mining company to develop the mine.

Gold is still here in Southern California. Geologists claim that over ninety-nine percent of the world's gold still remains to be recovered. I have found gold; others have found it. We are finding it today, and you can find it too. I have taken gold out of

A rich strike by a miner in the early 1900s.
(Photo courtesy of Title Insurance and Trust Company)

every area listed in this book, and I have had a lot of fun doing it. You will enjoy prospecting with your family or alone. (A note of caution: be extremely careful about open mine shafts and old tunnels. The shafts are sometimes overgrown and the tunnels can collapse.) You'll learn about the history of this region and about the people who made that history. You will see the hills, mountains, valleys, and land around you with new eyes. You will find beauty in the smallest pebble—and you might strike it rich while you are at it.

A second word of warning: once you find your first piece of gold, you are going to get gold fever. Don't worry; it's a pleasant disease. In fact, you will enjoy it.

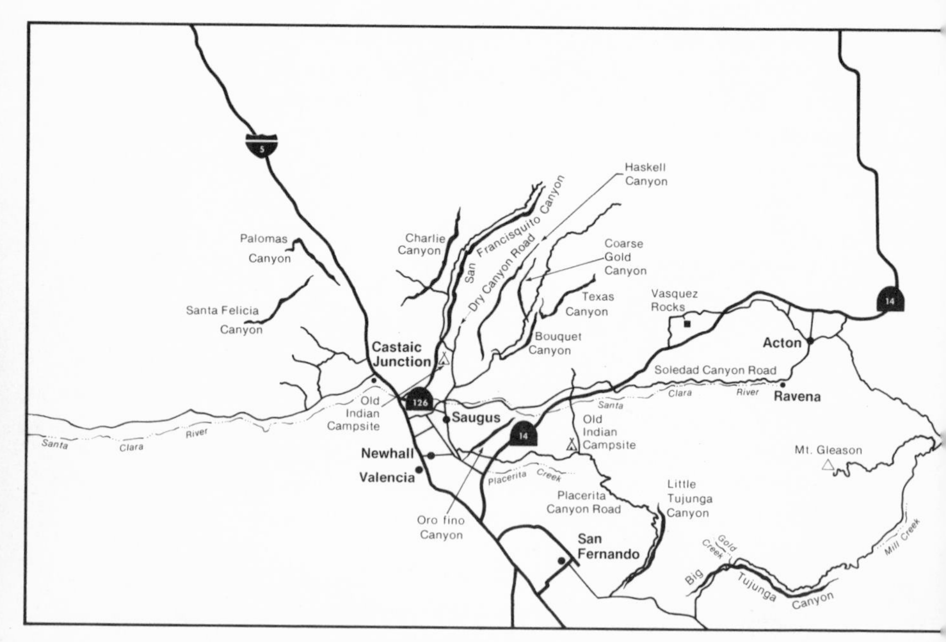

Newhall-Saugus and Acton Area

WHERE TO FIND GOLD

1 NEWHALL-SAUGUS

"In Placerita Canyon, March 1842, Francisco Lopez y Arballo, while gathering wild onions from around an old oak, discovered gold particles clinging to the roots of the bulbs. It is estimated that $80,000 in gold was recovered as a result of this discovery." Thus reads a California historical marker located in Newhall, California. The oak is known as the "Oak of the Golden Dream," and it is still standing today in Placerita Canyon State Park. Since I found my first piece of gold here (fifty feet from the park office and museum), I have selected it as the starting point to aid you in *your* golden quest.

Placerita Canyon

From Los Angeles, take the Golden State Freeway (Interstate 5) to the Antelope Valley Freeway (State Highway 14) to the Placerita Canyon offramp. Go right to the Placerita Canyon Park entrance. You will find water here for panning about seven months out of the year, usually from January through July. The rest of the year you will have to hike up the creek to find places where water is running.

You can also drive up Placerita Canyon Road, about a mile, to a locked gate on the right-hand side. You may park your car and hike down to the creek on a dirt road. You will find water here most of the year, and you are still in the Park.

A half mile or so from this area is the entrance to the Walker Ranch, currently used by the Boy Scouts for camping. This area is known as the Walker Ranch Placer Deposits. As you go downcreek, look on the left bank for tailings from the placer workings. According to Tom Walker, a grandson of the original Walker, over a million dollars in gold was mined from the hills and gullies around the present highway.

To reach Oro Fino Canyon, take Placerita Canyon Road, go under the Antelope Freeway Bridge, and turn right on the Sierra Highway to the top of the hill. Take the dirt road to the left and follow it down the hill into the Canyon. There is no water here.

The hills of Texas Canyon are composed of prehistoric river gravels. (Photo by James Klein)

San Francisquito Canyon

From Los Angeles, take the Golden State Freeway (Interstate 5) to the Valencia Road turnoff. Go east to Bouquet Canyon Road, then north to Seco Canyon Road. Follow it to the fork. Take the left fork to the National Forest boundary marker. About one mile from the marker, on a good paved road, you will find the best placer area. Unfortunately, you'll find water here only about six months out of the year.

Farther up the road, you will find water year round. Look up the small canyons running down to the road and you will see the remains of old diggings. At one time, over five thousand Chinese worked these deposits.

Treasure Tale

There are several treasure stories connected with the tragic collapse of the San Francisquito Dam thirty years ago. One of the most interesting is the story of a family of ranchers who lived in the canyon. They had a fine vineyard and winery, and marketed cattle and honey. They also mined the placer gold deposits. The head of the family was supposed to have hidden several thousand dollars in gold coins around the old adobe-brick house. After he died, the sons started tearing down the old adobe building and found many gold coins stuffed into plastered-over cracks. When the dam broke, it washed away the old building, scattering the remaining coins downstream, where they are still found today.

Point of Interest

At the mouth of the Canyon, across the creek from the old windmill and up the side of a hill, you will find an old cemetery. In it is the grave of a Mexican woman whose life spanned three centuries from 1796 to 1904.

Texas Canyon

Farther up Bouquet Canyon Road from San Francisquito Canyon is Texas Canyon. The turnoff to Texas Canyon is marked by a National Forest roadsign. The best placer areas are about two to three miles from the turnoff and past the place where the road crosses the creekbed. It's a dirt road but a good one, once you pass the homes. There is water for panning here only during the rainy season.

Point of Interest

As you travel up the Texas Canyon Road, you will come to a place on your right where you will see two old mine tunnels in a

The town of Newhall in 1885.
(Photo courtesy of Title Insurance and Trust Company)

hill. In the clearing, between the road and the hill, you will find the remains of an old camp. Carved in the trunk of the old giant oak tree, which stands in the center overlooking the campsite, is "1872." This marks the year of the gold discovery in the Canyon and the beginning of work on the deposits. You'll have to look up to see this date. The tree has grown during the past hundred years, and the numbers are now over seven feet above ground level.

Other Gold Locations

Off Bouquet Canyon Road, you'll reach several Canyons worth exploring.

Some of the best local placer mining was done on the hills running up to the junction of Dry Canyon and San Francisquito Canyon Roads. There are homes built on the hills now, but there are still places where you can recover gold. Try in spots where grading has been done and the gravel is exposed. There is no water here.

There is a good dirt road up Dry Canyon. The road ends at a locked gate and there is no water. Take side roads to the hills to find the old gravel deposits.

Haskell and Coarse Gold Canyons are off Bouquet Canyon Road and are marked by National Forest road signs.

A good dirt road leads you to some hills and gravel deposits in Haskell Canyon. I found a bottle here, in perfect condition, in the brush on the side of a hill, dated in the 1800s. No water is available.

Coarse Gold Canyon is explored by another good dirt road. There are gravel deposits here but no water.

To reach Charlie Canyon, take San Francisquito to the Y-fork where the road widens. Take the dirt road on the left. You'll find auriferous gravels here but no water.

San Felicia Canyon was the location of the first large gold camp in the area. The largest nugget (worth $1500 at the time) taken from the Newhall-Saugus area was found here. From the Golden State Freeway, take the Elizabeth Lake Road turnoff. Go west to the business road and then north to the end of the pavement. Go through the fence gate on the dirt road (being sure to close the gate, as there are horses grazing here). Follow Palomas Canyon Creek for about two miles to San Felicia Canyon Road on your left.

At the junction of Valencia Boulevard and Bouquet Canyon Road is the Santa Clara River. There is water for panning here all year. Bring your dirt from other areas and wash it here. The gold in this area will be very deep down, maybe twenty or thirty feet down.

Prospecting Tips

Practically all the canyons in this area have gold. The source of the gold is an ancient riverbed that eons ago must have been channeled to the sea. According to old reports filed by state geologists, wherever the present creeks and streams cut this ancient streambed and red dirt is exposed, the placering is good. The redness in the earth signifies the presence of iron oxide in quantity. Iron is sometimes referred to as the mother of gold, because the red dirt is the gold-bearing gravel. One last tip. One old-timer said to me once, "Yep, the red dirt's good all right, but the white sand under the red dirt is the richest of all."

Treasure Tales From Nearby Areas

Between the Newhall-Saugus area and the Acton District are the Vasquez Rocks, the locale of another treasure story. Tiburcio Vasquez was one of the most feared bandits to roam the Southern California countryside. After he was hung in 1875, there was much speculation about the quantity of spoils he must have accumulated during his infamous career. He used several places as hideouts, and all are thought to be potential sources of hidden treasure. The rocks just off Highway 14 between Newhall and Acton, named for the bandit, are said to have been one of his strongholds. Many people have searched here, but if they have found anything, they have kept it to themselves.

About two miles to the east are the Vasquez Caves, also named for the outlaw. Vasquez is known to have favored caves as a hiding place, because they offered him the greatest protection from posses. The caves, then, would probably be the best area to search.

The only generally-known treasure recovered so far was an iron kettle filled with copper coins dating from 1600 to 1777. The kettle was found buried about three feet deep in a dry creekbed in the Vasquez Rocks region. A human skull with a stone ax imbedded in it was found with the kettle, so the kettle and skull are probably remnants of an Indian-Spanish encounter, and not part of a Vasquez cache.

The outlaw no doubt did hide hard-to-dispose-of items such as jewelry and silverware, but no one knows where. Maybe you will be the one to find them. Whether or not you are lucky, this is a beautiful place to visit and worth a trip.

The outlaw Tiburcio Vasquez is believed to have buried several treasure caches before his hanging in 1875.
(Photo courtesy of Title Insurance and Trust Company)

Old hotel in Acton.
(Photo courtesy of Security Pacific National Bank)

2 ACTON

This is one of the largest gold districts, geographically, in Southern California. It stretches from the Mill Creek area high in the San Gabriel Mountains to Soledad Canyon near Vasquez Rocks. The first prospectors in the area probably drifted into the Soledad Canyon region after the placers in the Newhall-Saugus region began to play out in the 1840s. They may have been following the water. The Santa Clara River, in some parts of Soledad Canyon, has water almost all year. The canyons around Newhall-Saugus usually go dry during the summer and fall.

Not until 1861 was there any real rush to the area, and it was copper, not gold, that triggered it. The copper was first discovered in 1853, but nobody was interested in anything but gold in those days. Then, in 1861, the copper ores were relocated by some prospectors who then found gold and silver lodes nearby. The rush was on. The copper boom was short-lived, as the ores occurred in pockets and there were no veins to follow. The gold ores were first worked by hand by Mexicans laid off when the copper mines failed.

The Soledad Canyon area had several up-and-down times over the next few years. Around 1868, when there were so many men working there that a U.S. Post Office was opened, the town of Soledad City sprang up. The name was later changed to Ravena City, in honor of one of the first miners to work the area and to avoid conflict with the town of Soledad in Northern California. Ravena became a ghost town in 1877 when the miners moved a few miles east to the new railroad siding at Acton. The Soledad district then became known as the Cedar Mining District and, a few years later, the Acton District.

The most productive mine in the district was the Red Rover. First worked by the Mexicans, it was taken over by American miners around 1882. Other large producers were the New York and the Puritan. Roads in the area still carry the names of these mines from the old days. The New York mine is said to have produced over a million and a half dollars in gold during its lifetime. Both the New York (later known as the Governor) and the Red Rover were worked intermittently until the 1950s. The New York mine is said to be the most extensively worked gold mine in Los Angeles County. The remains of the Red Rover are easy to reach, and you should take a look at them so that you can see what a large-scale mining operation looked like.

Another mine that has been worked until recent times is the Monte Cristo near Mill Creek. Just last spring, I saw new claim stakes on Mill Creek just below the Monte Cristo claim. People have been doing a lot of work around the old mine. I don't know if they are working the old vein or following some new leads, as the area is heavily posted with No Trespassing signs, and I have never been able to find the people who own it. (I own a claim nearby—not the same mine I spoke of in the Introduction—that has assayed out at one-third ounce of gold per ton, which is pretty good.)

Nothing in Acton today gives a clue to its historic past as a boom town. Today, it's a nice quiet little town with a few stores and a post office. The Acton area is primarily a lode gold district with some placer gold coming out of Mill Creek and the Santa Clara River in the Soledad Canyon area.

Soledad Canyon

From Los Angeles, take the Golden State Freeway (Interstate 5) to the Antelope Valley Freeway (Highway 14). Take the Antelope Valley Freeway to the Soledad Canyon Road turnoff. Go right on Soledad Canyon Road. Placering has been done with some success in Soledad Canyon from Lyons to the site of Ravena, about three miles west of Acton.

Acton

From Los Angeles, take the Golden State Freeway (Interstate 5) to the Antelope Valley Freeway (Highway 14). Take the Antelope Valley Freeway to the Crown Valley Road turnoff and go right to the town of Acton. The mines were in the nearby canyons. Going left on the Crown Valley Road will take you to the Governor mine. To get to the Red Rover mine, go right on Escondido Canyon Road where it joins Crown Valley Road. Stay on it to Red Rover Mine Road, turn right, go under the freeway, and stay on it as it climbs up the canyon. The mine will be on your left on the side of the hill. Look for tailings and the headframe.

Treasure Tale

A prospector friend of mine, Darrel "Hardrock" McCaskell, told me that he has been on the trail of a fabulous lost mine in the Acton area for nearly a year. He first heard of the lost bonanza when he took some ore samples in for assaying. The assayer told him the following story.

Nearly forty years ago, a stranger brought in some material to him for assaying. The metallurgist's eyes nearly popped out when he looked at the rocks handed him. Here was the richest ore the assayer had ever seen. The rocks were laced with pure wire gold. The prospector was a poor man with no money to develop the mine, so the assayer offered to go into partnership with him. A deal was struck, but now comes the twist to the tale.

After he had outfitted the prospector, the metallurgist sent him back to bring out some more samples of the ore. About a week later, he learned that the man had been found dead, shot in the back, in a small canyon near the town of Acton. An old sheepherder who lived nearby was brought in by the police and questioned for several days. They could never prove that he had done it, even though he had threatened to shoot many people who had ventured into that area. He was not right in the head, they say, and felt that everyone was trespassing on his land, although he had no legal claim to the land, being a squatter. The assayer then became afraid to go look for the mine himself. Now, he was too old.

The assayer offered "Hardrock" a half-interest in the mine if he could relocate it for him. "Hardrock" says he has found the remains of the sheepherder's cabin, and he feels he is close to finding the rich lode.

The people around Acton are reluctant to discuss the incident, so you won't get much help there. Look in the small canyons for the old cabin. The prospector must have been heading for the lode when he was shot; maybe the sheepherder was not so crazy after all, and had something to protect. Perhaps you can find it before "Hardrock" does.

Prospecting Tips

The placers are not worked as much as in some other areas, except at Mill Creek which is worked quite a bit. The lode deposits are gold-bearing quartz in diorite and schist. The gold sometimes occurs with pyrites.

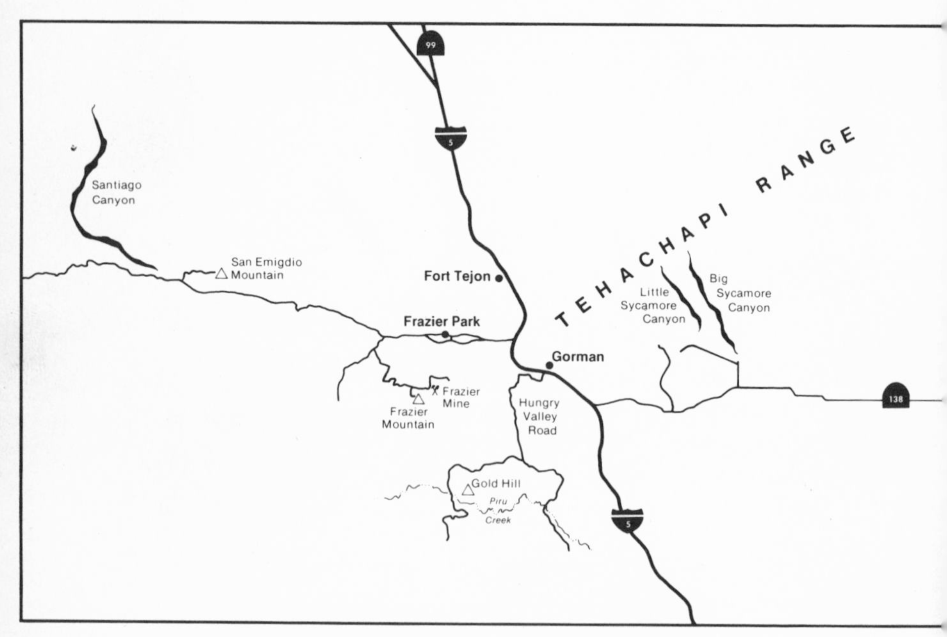

Frazier Mountain and Piru Creek Area

3 FRAZIER MOUNTAIN-PIRU CREEK

One of the West's most exciting lost mine tales concerns these mountains. The story of the Lost Padre Mine has, to my way of thinking, more truth in it than most other legends. Frazier Mountain and Piru are considered to be two separate mining districts. Since they are so close together and represent two different types of mining, I have combined them here. Frazier Mountain, primarily, has been a hard-rock mining district, with a minor amount of placering. The reverse is true of the Piru Creek area.

According to reports, placering was done here as early as 1841. The first known person to have mined the placers was Andrew Castillero, who is said to have brought out quite a bit of coarse gold from the gravels of Piru Creek. In 1842, the first gold was shipped from this area to the United States Mint in Philadelphia.

The Frazier Mountain-Piru area is one of the most beautiful areas in Southern California, and it has excellent camping facilities. You can spend a most pleasant and, with luck, profitable weekend prospecting here.

Piru Creek

From Los Angeles take the Golden State Freeway (Interstate 5) north to Gorman. Take the Hidden Valley turnoff. Go west on the Hidden Valley Road and follow the signs to the Gold Hill Guard Station. Continue on past the Guard Station to Hunters Overflow Camp Station. This is Piru Creek. The placers are in the creek gravels and in the older terrace deposits on the hills north of the creek. They run from this area west to the junction of Lockwood Creek with Piru Creek.

Frazier Mountain

Farther north on the Golden State Freeway (Interstate 5), you will come to a sign saying Mt. Pinos Recreation Area-Frazier Park. Take this road west until you come to a fork. Take the road to the left, and follow it for about a mile. Watch for a road on your left. This is the Frazier Mountain Road. The old Frazier mine was on the eastern slope of the mountain. It was discovered in 1865 and produced over a million dollars in gold.

Point of Interest

Be sure to visit old Fort Tejon State Park while you are here. It was built in 1854 for control of the Indians in this area, and to protect the ranchers in the valley. This was also the former home of the Army Camel Cavalry.

Treasure Tales

Fort Tejon is also a jumping-off point in our quest for the location of the Lost Padre mine. The tale begins back in the days of the missions. Soon after the San Fernando mission had been established, the fathers learned from the Indians of a large deposit of gold in the mountains behind Fort Tejon. The mission priests, with help of the Indians, worked the mine for several years. There are said to be records in Mexico City noting the amounts of gold shipped from the Mission San Fernando to the Mother Church in Mexico City. Even after the missions were forced to close, a priest was said to visit the Indians. He would perform all his clerical duties, and then he would take several of the young braves to the mine where they would gather all the gold ore that his pack mules could carry. Legend says that he made this trip to Mexico City each year for ten years. On his last trip to the mine, he was returning to Mexico City with ten mule-loads of ore when he was attacked by a band of renegade Indians and slain.

Nothing more was heard of the lost lode until an old prospector staggered into Fort Tejon one day. He was dragging a small sack filled with the richest ore ever seen by the men at the Fort. After he had been nursed back to health, he told of finding the remains of an old Spanish mine shaft with incredibly rich ore scattered around the site. He said that the mine was in the mountains west of the Fort, and that he had been lost for several days before he found his way to the fort. Unfortunately, when the men from Fort Tejon were leaving to go back to the mine site with the old prospector, he was thrown from his horse and killed.

Several years later, one of the Fort Tejon men was told of an old Indian who had worked with the padre at the lost mine. The Fort Tejon man, now a successful Bakersfield businessman, was able to persuade the Indian to lead him to the mine. When they had been out for two days, the old Indian had a dream in which the dead padre appeared to him. In the dream, the padre warned the Indian not to take the man to the mine. He was so frightened by the dream that he ran all the way back to Bakersfield. The strain was too much for the old Indian and he died. He had told no one where the mine was located.

The Lost Padre Mine is still lost. Maybe you will be the one who finds it. If you do, you will never have to work again. The mine is thought to have been in the San Emigdio Mountains. Try your luck at either San Emigdio or Santiago Canyons.

The lost mine of San Buenaventura Mission is thought to have been in the San Emigdio Mountains also.

Another lost mine story related to Fort Tejon concerns a young Indian and his wife. The couple had been thrown out of the tribe for some offense, and they lived in the hills around the Fort. In their wanderings, they discovered a rich ledge of gold in the Tehachapi Range near the Fort. For several years they traded their gold for food and supplies at the Fort, and then they simply disappeared. You might look in little Sycamore Canyon.

There are several other lost mine stories connected with the area. One of these is the tale told of a deer hunter who found a mine shaft with an iron door on the side of Frazier Mountain. He could never relocate the shaft when he returned with help to clear the debris surrounding the door. Some rich ore was said to have been found in the return search, and some crude mining equipment, but that is all.

It is possible that there is only one mine which has been rediscovered by different people at various times. The Iron Door Mine is also called the Los Padre Mine by some researchers. Could it also be the Lost Padre Mine? It is my opinion that the gold shipped by the priests at Mission San Fernando to Mexico came from the rich placers of the Newhall-Saugus area. These were much closer to the mission, and it is unlikely that experienced prospectors and miners would not have discovered them. If the Indians knew of the gold in the much more remote and difficult-to-reach San Emigdio Mountains, they would have been aware of the rich placers in the canyons.

This is not to say that there is not a very rich lost mine in the mountains north of Los Angeles. Too many stories exist to discount its existence. It's there somewhere; maybe you will be the one to find it.

Prospecting Tips

The placer deposits at both Frazier Mountain and Piru Creek are in the recent stream gravels and older terrace deposits above the present streams and creeks of the area. The lode deposits occur in quartz-bearing gold veins with pyrite. The Southern California Prospectors Club has a claim on Piru Creek about two miles north of Hard Luck Campground where they have been pulling out some pretty good gold.

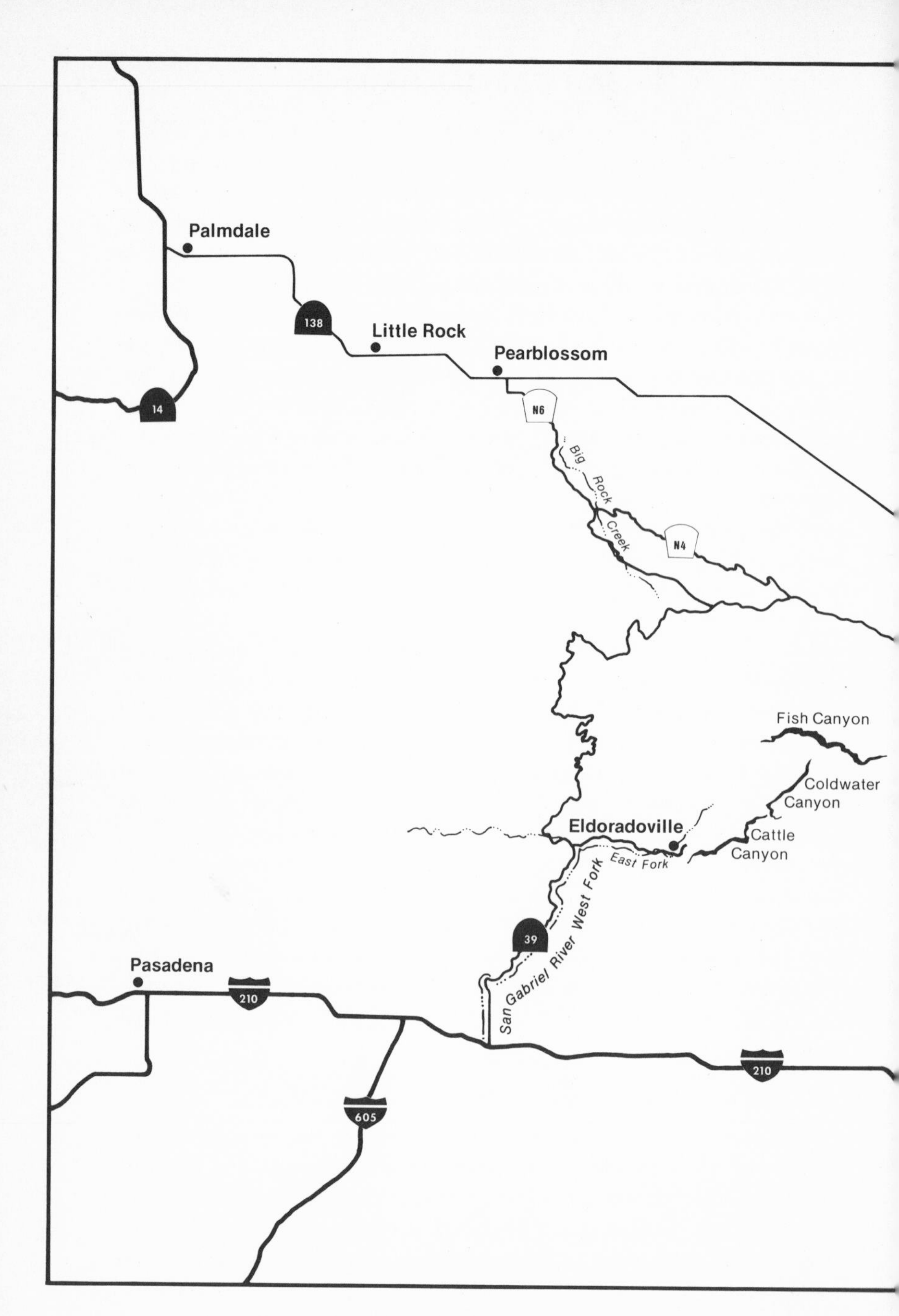

San Gabriel Canyon Area

4 SAN GABRIEL CANYON

This is the most popular area in Southern California for the weekend prospector. Millions of dollars in gold have been taken out of this beautiful canyon. The most productive area has been the east fork of the San Gabriel River, where gold has been taken from the gravel bars for a distance of over five miles.

Both Cattle and Coldwater Canyons have been a source of gold placers. The junction of Cattle Canyon with the east fork of the river was the site of Eldoradoville, which sprang up during the gold rush. During 1861, according to the Department of Mines and Geology, an average of twelve thousand dollars a month in gold was shipped by Wells Fargo Express from deposits in the Canyon.

A few years ago, a gentleman from Montebello made an agreement with a rock and sand company based at the mouth of the canyon to arrange a series of traps to find gold as the gravel was being processed. In order not to interfere with their normal operations, he would clean the traps out on the weekend. I am not sure how much gold has been recovered by this method.

Nothing remains of the original town of Eldoradoville today except an old mine shaft. The site of the town, though, is a pleasant picnic area. The shaft is a few hundred feet up the Glendora Mountain Road from its junction with the East Fork Road. In the shaft, one can find some interesting debris left from the minor gold rush of the thirties. Apparently, at one time this mine shaft was used as a dump, for thousands upon thousands of beer cans are deposited here. These must be some of the first beer cans ever manufactured, since the cans give opening instructions and make claims that would never be accepted today: "Refreshing, healthful, invigorating, non-fattening . . . adds zest to all meals!"

East Fork of the San Gabriel River

From Los Angeles, take any freeway to the San Gabriel River Freeway (Interstate 605). Take Interstate 605 north to the Foothill Freeway (Interstate 210). Go east on 210 to Azusa Boulevard (Highway 39). Go north on Highway 39, past two dams, up the canyon until you come to a large bridge. Turn right on East Fork Road, over the bridge. Look up the small canyons running down to the East Fork Road, and you will discover the old workings.

Miner's cabin.
(Photo courtesy of Title Insurance and Trust Company)

Quite a bit of terrace gravel remains. Follow this road until you pass the Hunters Camp Williams Campground and trailer park. Here, placering is done all along the river and on up for several miles.

Prospecting Tips

The river gravel is pretty well worked out in this area. You have to dig really deep to find anything worthwhile. But, if you are willing to dig down several feet, you will turn up some color. The best places are the old gravel benches high up on the canyon walls. There are several low-grade hardrock mines in the area. The veins are of gold-bearing quartz.

Fish Canyon

From Azusa Boulevard (Highway 39) take Foothill Boulevard (Highway 66) east to the first street on the right after you cross over the river. This is Fish Canyon Road. Drive up Fish Canyon Road as far as you can. The rest of the way you must go on foot. Half way up, the road changes to dirt, but it is still a good road.

Treasure Tales

One of the most often-told treasure tales of San Gabriel Canyon has to do with an old Indian woman sheepherder. One day, while rounding up her flock, she discovered an outcrop of rich, gold-bearing quartz frequently called jewelry rock. It was so rich that she was able to sell it to a Monrovia jeweler in chunks, right from the vein! The Indian woman is said to have kept her flock in Fish Canyon. The rich outcrop she discovered is most likely at the mouth of the Canyon. Look also in the small canyons and washes coming into Fish Canyon. Some tales say that this particular gold-bearing quartz was found in blue clay.

Another Fish Canyon story regarding lost treasure has to do with Fish Fork (off the east fork of the river). An old prospector and his mule were coming out of Fish Fork with a load of rich ore when the mule either died or was killed accidentally. The old miner was grief-stricken, as the mule had been his companion for many years. The old man removed as much ore as he could carry, then buried his old friend, the mule, and the rest of the ore. The prospector carved a cross on the rock walls marking the gravesite. Before he could return and dig up the rest of the ore, heavy rains hit the Canyon, causing a great deal of flooding. The violent floods carried away not only the body of the mule and the rich ore, but several feet of earth as well; for the carved cross, originally four feet from the ground, now stands nine feet from the ground.

To find the site, from the East Fork Station at the end of East Fork Road, take the fire road. You will have to hike. No cars are allowed beyond the gate to Fish Fork Campground. It is about five miles from the gate to the campground. Walk up Fish Fork until you see the cross on the wall of the canyon. From here to the junction of Fish Fork with the east fork of the San Gabriel River at the campground, will be the most likely place to search.

There is another treasure story dealing with the east fork. It was during the 1800s that four men held up a stage down in the San Gabriel Valley. They fled into the mountains with their loot, then split up. All were run down and shot by a posse, but none of the twenty-five thousand dollars taken in the robbery was found on the bandits. Most of the searching has been done around nearby Camp Oak Grove. The bandits were on the run and couldn't have buried the money too deep. Look for a prominent landmark, because the bandits would have needed one to guide them in relocating their treasure.

There are some nice campgrounds on the east fork and the river is stocked with trout each week. Bring your family for a pleasant weekend of prospecting, treasure hunting, camping, and fishing.

Down in the San Gabriel Valley there are several tales of lost treasure you may want to investigate. One of the most-told tales is of the Tiburcio Tapia Treasure. Don Tiburcio Tapia was somewhat of an *alcalde* of the area around Cucamonga during the early history of the Valley. He is thought to have hidden several caches of gold during his lifetime; the biggest horde was hidden just before his death.

The advance of American troops from the south during the war for California's independence from Mexico caused the residents of the Valley to fear for their wealth. Several of the families entrusted their gold and jewels to Tapia for safekeeping. The old man took the treasure and buried it somewhere along the Arroyo Seco. Two days later, Tapia was stricken with a heart attack and died without revealing the exact location of the treasure. An Indian servant helped him bury the valuables, but was made to take a vow of secrecy by his master which he was never to break. All that is known is that the gold and jewels were buried under a large oak or sycamore tree. The area around the Arroyo Seco is now built up. This treasure will be lost, most likely, for a long time.

There is a ghostly aftermath to the Tapia tales. Years after Don Tapia's death, his daughter and her husband were living in the old adobe house where he had died. One night, the young woman was awakened by a voice calling her name. The frightened woman searched the room for the source of the voice. She could see nothing but a bright glow coming from a spot on the wall across from the bed. She woke her husband, told him of the voice, and showed him the light coming from the wall. To reassure her, he got out of bed, took his knife, and plunged it into the glowing wall. Instantly, the light went out, the wall gave way under the knife, and the clink of metal striking metal was heard. The couple stared at each other in amazement. The walls were supposed to be solid adobe. The knife had made a hole big enough for a fist to go through where the ghostly light had been. When the son-in-law reached his hand into the hole to see what the blade had struck, he discovered a golden horde. Don Tapia must have hidden the coins there. For years after, people poked holes in the walls of the adobe seeking other hiding places, but no more were found. Did the ghost of Tiburcio Tapia return to reveal one of his hidden caches to his daughter? Who knows?

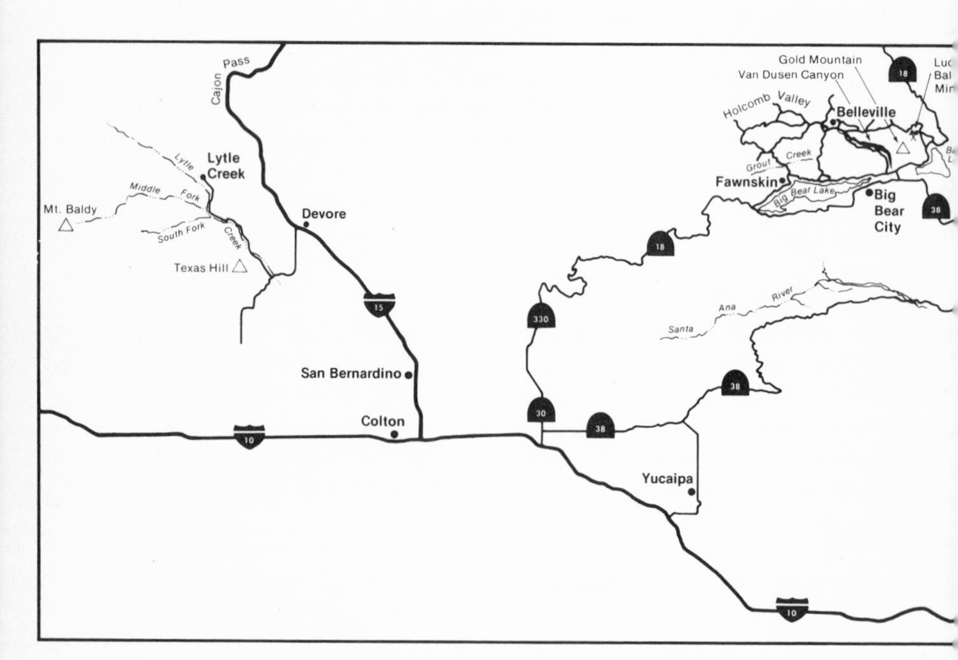

Lytle Creek and Holcomb Valley Area

5 LYTLE CREEK

Lytle Creek has been the scene of considerable gold mining excitement in the past. Named for a Mormon, Andrew Lytle, who settled in San Bernardino in 1851, it has produced over a million dollars in gold, according to reports. The gold was all placer, being found in both the recent stream gravels of the creek and in the older terrace deposits on both sides of the creekbed. At one time, there were claims running all the way from the mouth of the canyon to the headwaters of the creek high on the east slope of Mount Baldy.

The first gold was discovered in the creek gravels in May of 1864. One of the first men to mine the deposits was a man known as Abbot. Abbot later killed a man named Keir, who Abbot said had tried to jump his claim.

The placers were predominately hand-worked until 1867, when the Hardpending Company, under the direction of a Mr. Winder, began hydraulic operations in the canyon. The hydraulic operations they worked were in an area called Texas Point. Look on your left as you enter the mouth of the canyon. You can still see the scars left by the giant hoses. Hydraulic mining continued until almost 1880, with reports of two thousand dollars a week in gold being recovered from the gravels. A dispute with the placer miners of the Creek over the large amount of water used by the company may have caused its demise. At least one killing is said to have occurred during the dispute.

There are many pleasant campgrounds in this area and one can spend an enjoyable weekend prospecting here. The most productive area was the North Fork. Water can be found all year at the forks.

Lytle Creek

It is relatively close. From Los Angeles, take the San Bernardino Freeway (Interstate 10) east to San Bernardino. Take the Barstow Freeway north to the Devore Road turnoff. Follow Devore Road west to Lytle Creek Road. Go right on Lytle Creek Road to the mouth of the canyon. From here on up are the gold-bearing areas.

Point of Interest

The first house in San Bernardino, and also a Morman fort, stood at the corner of Arrowhead and Court Streets near Starke's

Hostelry. Out of town, the Cajon Pass was the western extension of the famed Sante Fe Trail. Look for two historical markers in the Pass noting this.

Prospecting Tips

There was an item in an old newspaper clipping years ago that someone had found a rich lode in the Canyon. It was said to have contained gold, silver, and platinum in large quantity. What happened to the man or the mine, I don't know.

I *have* seen a piece of ore displayed at Keene Engineering that assayed out at thirty-five thousand dollars a ton. It was rust colored with some grey, and it looked somewhat like slag. This good ore came from Lytle Creek.

The gravels on the walls of the canyon will hold the most gold. There is a theory that a large river ran from the north clear to the Gulf of Mexico and flowed over what are now the Sierra Nevada, San Gabriel, and San Bernardino Mountains. Recently, a man claimed to have traced this ancient riverbed into the nearby Cajon Pass. The ancient river is called the Big Blue, from the color of the quartz gravels containing the gold found in it. So far, only about sixty miles of it have been discovered and worked, and that sixty miles has paid over three million dollars a mile. If you can locate a portion of it, you will indeed be fortunate.

Treasure Tales From Nearby Areas

There is a tale told of bandit loot hidden nearby. In the late 1800s, some men robbed a stagecoach along the old road that ran beside the Santa Ana River in Colton. They fled with their booty up Lytle Creek Wash and hid in the foothills. There, they were captured; but the gold was never recovered.

During the 1870s and 1880s, Starke's Hostelry at the corner of Third and Arrowhead in the city of San Bernardino was known as the miners' rendezvous. Prospectors and miners from Holcomb Valley, Randsburg, Dale, and Lytle Creek came to Starke's for a few days of drinking and gambling, or as long as their gold held out. There was a big backyard behind the hotel that was used for hanging the wash, among other things. One miner is said to have been on a hot streak one night and won nearly twenty thousand dollars in gold. That night, when he was awakened by a robber searching his room, he grabbed his pistol and started shooting; so did the burglar. Both men were killed and the gold was never found. Most people figure that the prospector had buried it

Hydraulic mining operation cutting down the hillside at Lytle Creek. (Photo courtesy of Title Insurance and Trust Company)

somewhere in the backyard when he went to the outhouse before going to bed.

There's a story that has gone around for a long time that a hiker walking up a wash in these foothills found a large raw ruby. Another story has an ancient ruby mine being found in this region. It might pay you to pick up and check any red rocks you come across when prospecting here.

E. J. "Lucky" Baldwin's Doble Mine building in Holcomb Valley. (Photo courtesy of Security Pacific National Bank)

6 HOLCOMB VALLEY

Girls, gambling, guns, guts, and gold. Holcomb Valley had them all. No other area in Southern California typified the incredible cycle of boom and bust of a gold rush as well as this peaceful and beautiful Valley. Cradled high in the San Bernardino Mountains and reached by dirt road only, it is hard to visualize the Valley swarming with thousands of prospectors. Yet, after the first gold in the Valley was discovered by Bill Holcomb in May, 1860, so many people came flocking that three towns sprang up overnight.

The first town was Belleville, which had the first post office in the mountains. In 1861, Belleville was the largest voting township in San Bernardino County and almost became the county seat in 1862, failing by only a few votes. Clapboard Town and Union Town also sprang up during the height of the rush. Today, there is no evidence of these towns except rotting wood and rusting metal scattered around the sites.

The gold here is very fine, and you will have to be careful when you pan or you will lose it. There is a good campground right in the area of the deposits. This can be one of the most pleasant prospecting trips for a family to take.

Holcomb Valley

From Los Angeles, take either the Pomona (State Highway 60), San Bernardino (Interstate 10) or the Riverside (Highway 91) Freeways to the Barstow Freeway (Interstate 15). Go north to the turnoff for the mountain resorts (Highways 18 and 30). Stay on Highway 30, after Highway 18 heads off to Lake Arrowhead, all the way to Running Springs. At Running Springs you will pick up Highway 18 once again. Continue eastbound on Highway 18 to the Highway 38 junction. Take Highway 38 through the town of Fawnskin to the Holcomb Valley Road (National Forest Road Number 2N09), which will be on your left and marked by a white post saying Holcomb Valley Road. This is a good dirt road. Stay on Holcomb Valley Road until it dead ends at the National Forest Road 3N08. This is where the old workings are. The gold is found on both sides of the Holcomb Valley Road and, ahead and across National Forest Road 3N08.

Go east on National Forest Road 3N08 for about one mile from its junction with Holcomb Valley Road, and you will come to the meadow where Wild Bill Holcomb found the first gold. Nearby lies the site of Belleville.

Another three miles farther east on National Forest Road 3N08 is the site of the old Lucky Baldwin or Doble Mine. There are still some remains to be seen from its glory days. This is considered to be a district separate from the Holcomb Valley area. Gold mining was done here as early as 1800 by the Mexicans. It is primarily a hardrock mining area, but there are placer deposits located here as well. The old mining town of Doble was below the Doble Mine, and like Belleville, Clapboard, and Union, has disappeared.

Going west on National Forest Road 3N08 from the junction of the Holcomb Valley Road for about four miles will bring you to the Greenlead Mine area. This is on your right as you travel west. A word of caution. There are several shafts in the area and old mines are dangerous. Unfortunately, every so often someone has to be rescued from a fall down a mineshaft or from being trapped in a collapsed tunnel. Even throwing rocks down a shaft can result in the edge you are standing on crumbling and taking you along with it. A small rock *thrown from a safe distance* will tell you how deep a shaft is.

Treasure Tale

Holcomb Valley, like all gold districts, has its own lost mine story. This one goes back to the days of the Valley's gold rush. Two men (one named Van Dusen; the other's name has been lost to time) were working a mine in one of the canyons leading into the Valley. According to the other miners, their ore was very rich. They never spent more than they needed, nor did that take any ore to San Bernardino to bank. They were assumed to have accumulated quite a large horde. One day, the partner with the forgotten-name was seen leading their mules full-packed out of the hills. After a week had gone by and there was no sign of Van Dusen, several of the miners went up the canyon to check. They found Van Dusen dead. Shot in the back. The partner was never seen again. Many of the men tried to find the mine, but it was never located.

Years later, two prospectors were said to have worked the area and brought out some rich ore. These two miners also disappeared. A search was made for the mine again, but to no avail.

To try to locate the lost Van Dusen Mine, take Highway 38 past the Holcomb Valley Road. Turn left on Van Dusen Canyon Road (National Forest Road 3N09). The lost mine is said to be located in this canyon. Look for gold-bearing quartz in granite.

The remains of an old Spanish fort are said to be a few miles above Fawnskin in the Grout Creek area. This would be a good site for relic treasure hunters. You will have to search well because all that remains are the wall rock humps. From Highway 38, in Fawnskin, take National Forest Road 2N13 to Road 2N68Y. Go left on this road for a quarter of a mile. Here, you will take National Forest Road 2N68 or Grout Creek Road on your left. The old Spanish Fort is said to be in this area.

Point of Interest

There are several interesting things to see while in Holcomb Valley. Go to the Forest Service or to a real estate agent in Big Bear and get a free map of the area. It will show the location of Wilber's grave, the Hangman's Tree, Pygmy Cabin, Two Gun Bill's Saloon, and other landmarks.

Treasure Tales From Nearby Areas

The Lost Lee Lode is said to be located somewhere east of Big Bear. Lee was a miner who used to come into San Bernardino with rich gold ore to purchase supplies.

R.W. Waterman, who later became governor of California and who had mining interests all over, put up a reward for anyone who could find the Lee Mine. Apparently, the reward was never claimed. Lee's mine was without doubt rich. In 1879, he filed his claim stating that it was five miles northeast of Big Bear Valley. Many people testified that they had seen the rich, gold-bearing quartz Lee brought in for trade. The mine is lost, because Lee was shot and killed outside of town one day as he was returning to his mine.

Some will tell you that the famous Rose Mine discovered later was probably Lee's Lode. Nobody knows. Since his gold was still in the matrix, he must have been crushing it to separate the paying ore from the waste rock. Look for quartz tailings, and an *arrastre* used for crushing and separating the gold.

The Lost Macfadden Treasure

Bernarr Macfadden was one of America's first health enthusiasts. The wealthy magazine publisher and businessman is, interestingly enough, said to have buried large sums of money in several places across the country. In her book, *Barefoot in Eden,* Macfadden's widow told how she came across him burying a tin box full of money under a pine tree near the Arrowhead Springs Hotel which they owned. A box like the one she saw him bury was turned up by a bulldozer near his home in Far Rockaway, New York. It contained $200,000.

The arrastre, *or primitive quartz mill, was used by miners to crush gold-bearing quartz to powder in order to separate the gold.*
(Photo courtesy of Security Pacific National Bank)

Lost Mine of Yucaipa

At the foot of the San Bernardino Mountains near the city of Redlands, sits the little town of Yucaipa. For years, natives have told of people finding pieces of rich float (gold ore) in a wash outside of town. It is said to come from an old Spanish or even an early Aztec mine located in the foothills near the town. Some even say the mine dump has been found, but the tunnel still remains lost.

Diamonds in San Bernardino

If there can be rubies by Lytle Creek, you can guess that someone will be claiming that there are diamonds on this side of the Cajon Pass. A man is said to have recovered quite a few diamonds just a mile from the outskirts of the city of San Bernardino. Many miners feel that there is a very good possibility that there is a diamond field near San Bernardino. Keep your eyes to the ground, and maybe you will be the one to make the discovery. For years I have wondered if those bright shiny stones I find in my pan or sluice box have been pieces of clear quartz or diamonds. I *hope* they weren't diamonds.

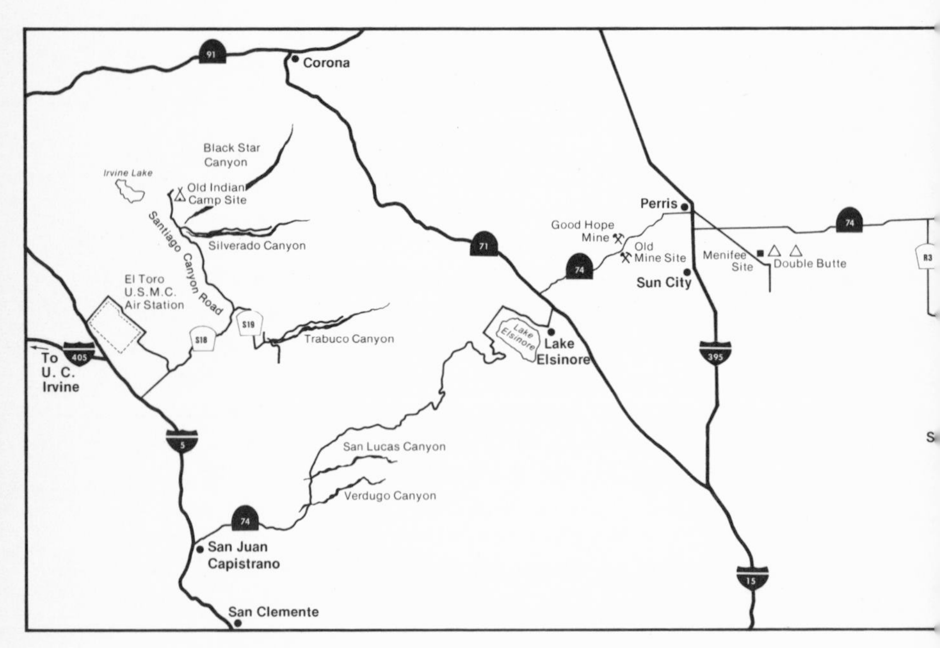

Pinacate-Menifee and Orange County Area

7 PINACATE-MENIFEE

Few people are aware that the mountains south of Corona contain an extensive gold belt. Pinacate and Menifee are listed as two separate districts by the State Department of Mines and Geology. For our purposes, we will combine the two.

Mining was done here as early as the early 1800s. The placers were worked rather extensively in the 1850s. The richest mine in the area was the Good Hope Mine in the Pinacate district, discovered in 1874. Large-scale mining was done here until 1903. Additional work was done here in the 1930s, when, during the Depression years, many of the original gold-bearing areas of California were reworked. Many men were more willing to take their chances in the hills than in the bread lines.

The remains of the huge operations at the Good Hope Mine can still be seen today. The main building, where nearly two million dollars in ore was processed, is still standing. There are remains of other old buildings in this area as well.

As for the town of Menifee, nothing remains. In my search for Menifee, the only clue I had was that it had been eight miles south of the town of Perris. Except for topographical maps printed by the United States Department of the Interior Geological Survey, no current maps show it. However, there is a railroad siding sign marking the site. According to one old-timer, "It was a pretty little town, sitting up against the mountain." This would put it somewhat southeast of the railroad marker. The mines were in these mountains, and the placer workings were in the hills south of the Good Hope mine. Today, there is some work being done by weekend prospectors. This area is not as overworked as some of the other gold-bearing areas in Southern California.

Pinacate

From Los Angeles, take the Santa Ana Freeway (Interstate 5) south to the Riverside Freeway (Highway 91). Go east on the Riverside Freeway to Corona. Take Highway 71 south out of Corona to Highway 74 (Perris Road). The gold is in the hills between Lake Elsinore and Perris.

The remains of the Good Hope Mine are on the left of Highway 74. Look sharp or you will miss it. The main building is still standing. Also visible from the same highway are the remains of a mine processing plant. A shaft is to the left of the plant and up the hill. Be careful, as the shaft is deep, flooded, and open.

The shell that remains of the Good Hope Mine in the Pinacate District. (Photo by James Klein)

Menifee

Follow the directions to Pinacate. Continue east on Highway 74 through Perris. Outside of Perris, Highway 74 crosses the Southern Pacific Railroad tracks. A dirt road runs alongside the tracks. Follow the dirt road till you come to a railroad siding sign saying Menifee. The town is set up against the mountain on your left. Some foundations can be found in the field. The mines were on the side of the mountain. If you like to collect old, large railroad spikes, quite a few are laying around this area.

Treasure Tale

Somewhat south of this region is the little town of Sage. It was here that Juan Chavez, a member of the Joaquin Murrieta band, lost two hundred thousand dollars in gold. While Chavez and some of the Murrieta gang were sleeping, one of the bandits took their gold. He buried it, thinking he would come back and get it later. The other bandits, however, caught and killed him. They made one mistake. They forgot to find out where he buried the gold before dispatching him. The treasure is still there, probably under some large tree or other prominent landmark.

Point of Interest

Highway 71 follows the route of the old Temescal Road. Watch for several State Historical markers along the way.

Lake Elsinore has several nice campgrounds for your weekend enjoyment. There are motorboat races on the Lake during the year, and also sky diving and hang gliding.

Prospecting Tips

Placering has been done with some success recently in the hills. Most of the prospectors I know who are working in the area are pretty secretive about how they are doing. This is a good sign that it's paying pretty well. The lodes are gold in quartz occuring in veins and seams with sulfides. The ore is fairly rich, running one-half to an ounce per ton.

Early photo of Mission San Juan Capistrano showing the exposed adobe walls. (Photo courtesy of Security Pacific National Bank)

8 ORANGE COUNTY

Not much gold has made its way out of Orange County. The biggest rush was for silver in 1878. It was a typical boom and only lasted for three years, ending in 1881. The location was Silverado Canyon. A State of California historical plaque marks the area today.

I am sure of two placering areas; one is a hardrock site I have heard about, and the other one I have worked. The hardrock site was originally mined by San Juan Capistrano Indians, who also worked placers in San Lucas Canyon. A lost mine story associated with the mission Indians has it that this particular mine had a gold-bearing quartz vein and had been worked for several hundred feet. A prospector friend of mine told me that a friend of his had made a strike in Verdugo Canyon and was filing a claim on it. The only place that has produced any gold for me, however, has been in the remains of an ancient riverbed high on the mountains overlooking Cañada de Los Alisos.

Trabuco Canyon

From Los Angeles, take either the Santa Ana (Interstate 5) or the San Diego (Interstate 405) Freeways south to El Toro Road. Go east on El Toro Road to its end. This is Trabuco Canyon. There is a nice campground here (O'Neil Park). Use this as a base for your prospecting.

San Lucas Canyon

Further south after El Toro Road on the San Diego Freeway is San Juan Creek Road (Highway 74). Take Highway 74 east for about nine miles. Look on your right for a sign saying, San Lucas Canyon Quarry.

If you can, try to get a key from someone in the Forest Service or the Sheriff's Department. When I was last up there, I could still see the old sluice boxes left behind. At one of the old diggings, there were chairs and a table set up and waiting for the miners to come in from work.

Verdugo Canyon is back about a mile and a half.

Treasure Tales

There are several tales of buried treasure here associated with the San Juan Capistrano Mission. The old mission was originally several miles east of its present site and near San Lucas Canyon. (San Lucas Canyon is shown as Lucas Canyon on some maps, incidentally.)

One treasure is said to consist of several bags of gold buried by a Forty-Niner going home with the fruits of his labor. He became ill and stopped at the mission to seek medical aid. After he had been there a few days, he decided to bury his gold for safekeeping. He chose the ruins of the old mission to hide it in, because it would be easy to relocate. He died, however, before he could recover his gold.

Another tale is of the padres being attacked by Indians and being forced to flee to the mission at San Diego for safety. They are said to have buried their church treasures under the foundations of the original mission. It is located somewhere in the mountains behind the mission.

Also of interest is a lost gold mine tale. The mine was worked by the Indians for the padres. The earthquake that destroyed the first mission also caved in the mine. Its location was lost from that point in time until the present.

Prospecting Tips

Trabuco is the only designated gold district in the county, according to the state Department of Mines and Geology. The lode gold occurs in narrow veins with tin, copper, zinc, and silver. The placer gold I have found consists of fine pieces which can be picked out of the pan only by using tweezers.

Treasure Tales From Nearby Areas

Juan Flores was a bandit who roamed the Southern California countryside, killing and looting during the 1850s. In 1857, Flores and his band raided the town of San Juan Capistrano, killing a shopkeeper. They took almost everything of value from the people there, and escaped into the Santa Ana Mountains with the booty. The sheriff of Los Angeles, John Barton, with a posse of five men, tracked the bandits to an area that is now known as Barton Mound. There, Flores and his band ambushed Sheriff Barton and the posse, killing the Sheriff and three of his men. Two members of the posse managed to escape and make their way back to Los Angeles to tell of the slaughter.

Don Andres Pico with a detachment of California Lancers rode out after Flores. They killed the bandit leader and twenty of his men. The loot from the raid, however, was never recovered. Since

Don Andres Pico, captor of Juan Flores.
(Photo courtesy of Title Insurance and Trust Company)

Barton Mound was a favorite hiding place for the bandits, most treasure hunters feel the loot is hidden somewhere in that locale.

Another lost treasure cache attributed to Juan Flores, said to contain forty thousand dollars, is supposed to be hidden in some caves behind Lake Irvine. Local residents up that way seem to feel that the treasure is still there. (When I was searching for it, by the way, I was warned that a mountain lion roamed the area.)

The Lost Treasure of Irvine Ranch

A stagecoach was held up by five men in 1868 near Corona del Mar. A posse from the city of Santa Ana came upon the men the next morning; all had been shot while still in their bedrolls. The money from the holdup was never found. Since the bandits had not traveled far, the gold might have been hidden close by. The location of the holdup is now the campus of the University of California at Irvine.

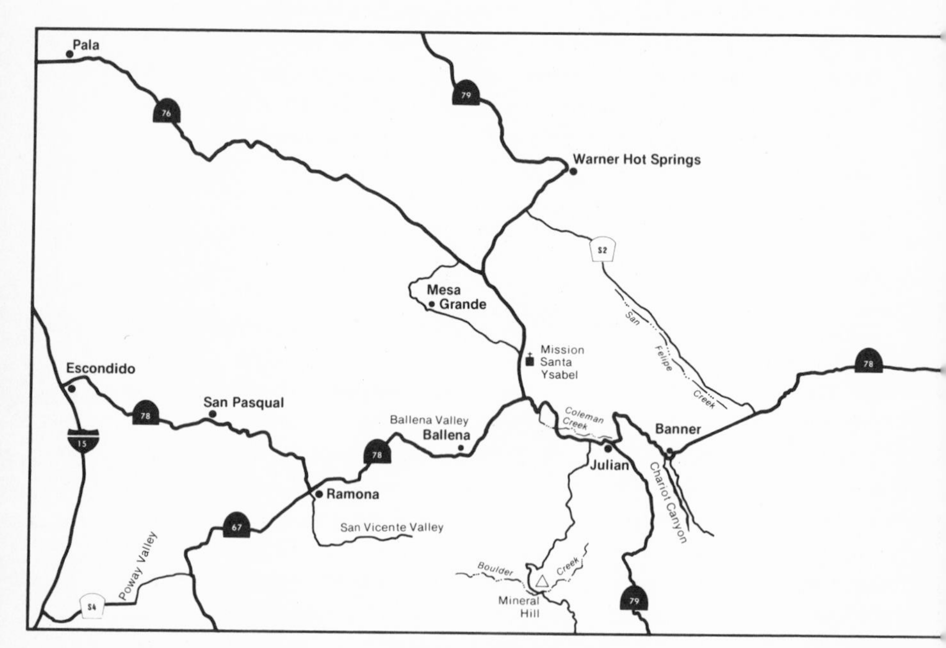

Julian-Banner Area

9 JULIAN-BANNER

Although it is nearly a two-hour drive from Los Angeles to Julian, it is such an extraordinary trip that it must be included in any book dealing with gold in Southern California.

The gold rush to Julian began when small amounts of placer gold were found by E. Wood and A.H. Coleman in the stream bed of Coleman Creek in January, 1869. This discovery started the search for the source of the gold by veteran prospectors from the Mother Lode diggings of Northern California. The placers of the Mother Lode were now played out, and the mining had been taken over by large companies. Julian became a mecca for the independent prospectors who swarmed over its hills, staking out claims, and developing the mines that created the towns of Banner and Julian.

Today, just one store marks the site of Banner. The site is five miles east of Julian on Highway 78. Mines dotted the mountains north and west of town. Julian, however, continues to survive as a strong agricultural community.

Prior to the discoveries at Julian, there had been considerable gold mining around the town of Escondido. Equipment used by the Escondido Mining Company, including its stamp mill, was moved to Julian during the gold rush.

Although this area is primarily a lode-gold region, the Indians, and later the Mexicans, had been mining the placers of Ballena Valley for many years prior to the discovery at Coleman Creek. The placer deposits here are unusual, in that, due to faulting and uplifting, fragments of an ancient riverbed lay exposed only on the tops of the present mountains. I found a reference to these deposits in a report from a field representative of the State Department of Mines in 1893. In his report, he states only that placering had been done on Ballena Mountain. *Ballena* is the Spanish word for whale and the mountain is shaped like a whale's back. (The best place and time to see the whale shape is one-half mile down old Julian Road in the late afternoon.) In prospecting the area, I have found the ancient riverbed exposed on other peaks in the area.

Julian

From Los Angeles, take any freeway to the San Diego Freeway (Interstate 405), go south on the San Diego Freeway to the city of

Indian graveyard at Santa Ysabel.
(Photo courtesy of Security Pacific National Bank)

Oceanside. Take Highway 78 east to Julian, using the Escondido turnoff.

The mines were in the hills north of town. For a fee, you can take a tour of one of the old mines. Julian is one of the most charming places in Southern California.

The site of Banner is a few miles farther east on the same highway.

Treasure Tales

Just a mile north of its junction with Highway 78, Highway 79 passes Mission Santa Ysabel. Santa Ysabel was not a true mission but an asistencia, or sub-mission. It was built by the padres of the San Diego Mission to serve the Yuman Indians of the Santa Ysabel Valley.

Two lost treasures are connected with the Mission. The first is said to contain nearly five million dollars in gold collected from the early California missions. The gold was brought here and hidden to protect it from the Mexican government when the government ordered the Franciscan priests to return to Spain. The padres, fearing that the Mexican government would seize the church's valuables if they tried to take them, decided to bury the valuables until the ban was lifted and they could return to their missions. The hiding of the treasure was entrusted to a single priest and a few trusted Mexicans. They buried it in the nearby foothills and started for the mission at San Diego. About halfway there, they were set upon by a band of renegade Indians and killed. Since they were the only ones who knew the exact spot where the treasure was buried, the secret died with them. Some people feel that the gold was hidden under the altar of the first church. The foundation and floor of the first church were found in 1963 and are still under study.

To add to the confusion, there is a large stone with a cross carved on it located near the present chapel. This is said to be one of three such rocks left by the padre to mark the site of the treasure.

The second lost treasure of Mission Santa Ysabel dates from 1851. Three men who had struck it rich in the Gold Rush up north were returning home with their gold. As they stopped to camp one night near the mission, the Indians kept them under observation. Unexpectedly, the three began to quarrel over the contents of a large deerskin bag. One of the trio killed the other two and started to flee with the bag. The Indians, thinking the bag must

General Stephen Kearny.
(Photo courtesy of Security Pacific Bank)

contain some great treasure, set upon the man, killed him, and took the bag to their chief. When the chief opened the bag, said to be as large as a man's head, he saw that it was full of gold dust and nuggets. Fearing the white man's revenge over the loss of so much gold, and thinking others might know of it and be following, he took the bag and buried it. He placed a curse on the gold and told his people that anyone who ever touched it or revealed its location would die a horrible death. (There is water in the creek south of the mission part of the year for potential treasure seekers.)

Points of Interest

On your way south, you may want to swing off the freeway a moment and view the former Western White House in San Clemente. It's the last exit as you go through San Clemente.

A few miles east of Escondido on Highway 78 is the site of the Battle of San Pasqual. Here, on December 6, 1846, a detachment of the First U.S. Dragoons, under General Stephen Kearny with Kit Carson as a scout, engaged a group of native California Lancers, commanded by General Andres Pico. Heavy losses were sustained by the American forces. This was one of the most important battles of the 1846-1848 Mexican War. The First Dragoons were encamped here for several days. I was told that the white crosses in the little graveyard here are for the American men who were killed in the fighting.

A little farther east on Highway 78, at the mouth of the canyon, is the site of the San Pasqual Indian village. The main part of the village was at the Y-junction of the creek.

You will also find the San Diego Zoo's Wild Animal Park nearby.

Right before you come to Julian, there are a group of ranches that are a treat to visit in the spring and in the fall. In the spring, lilacs are in bloom and in the fall, apples stud the trees.

Prospecting Tips

This is mainly a hard-rock mining area with a small amount of placer mining. The gold occurs in quartz with pyrites. An old-time prospector, with nearly sixty years of roaming these hills, showed me a pretty piece of gold-bearing quartz he found near Julian. The quartz is a steel blue-gray with tiny white specks in it. He says that this is the kind of outcrop to look for. Another old-timer I know claims he makes a living working with a dry washer in the Laguna Mountains.

The placers are in the present creek beds and in ancient creek gravels high on the mountains. I swear, and some are sure to disbelieve, that I got a pretty piece of gold out of a sandstone road cut on the new Julian highway.

There are several gold districts in this region. They are the Cuyamaca, Boulder Creek, Deer Park, Laguna Mountains, Mesa Grande, and Montezuma districts. Most of these districts had one or two good paying mines with a few short-lived strikes nearby.

To reach the San Vicente Valley placer deposits, travel through the town of Ramona on Highway 78 to the first stop sign at Main Street. The highway turns left here, but you turn right and make a quick left turn on the first street to your left. Follow it about one mile past Wildcat Canyon Road. The area between this road and the old Julian Highway is a region of projected placer deposits according to the State Department of Mines and Geology. There is water here only when it rains.

The old mining district of Mesa Grande had a boom that ran from 1880 to 1896. The mines were northeast of the town. The Mesa Grande Road is to the west off Highway 79, just north of Mission Santa Ysabel.

Gold was discovered in the Boulder Creek Mining District around 1885. There is still some small scale mining going on here. Boulder Creek runs south out of the Pine Hills area near Julian. The best place to prospect would be around Mineral Hill, where most of the old mines were located.

The whole area from Ramona to Pala and east to the desert is rich in gemstones. Some of the finest tourmaline you'll ever want to see can be picked up in the hills. Here I found a beautiful piece of black tourmaline sticking out of rock that I keep on my desk.

There is a lost beryl mine south of Banner. Miners were throwing the beryl stones into their dump at their diggings. They didn't know the stones' value at the time and could not locate the prospect hole later.

Treasure Tales From Nearby Areas

San Diego Mission Treasure

From down in the Poway Valley, comes another treasure story. This treasure came from the Mission at San Diego. According to this story, the padres feared that the Mexican soldiers, who had taken control of San Diego, would take their beautiful gold chalices and other altar ornaments. They entrusted the treasure to some church members who were loyal to the Mission. Just moments before the troops arrived, these trusted men slipped out the rear of the mission with their precious cargo. The padres delayed the soldiers as long as they could. The men managed to reach a point near Black Mountain where they buried the valuables and placed a large round rock over the spot.

Lost Ulloa Treasure of Oceanside

A quarter of a million dollars of this treasure is said to have been recovered. Since there was ten million in gold lost, it is still worth a look-see. The gold is part of the wealth taken from the Aztecs by Cortez.

Cortez commissioned one of his captains, Francisco de Ulloa, to search Alta California for the fabulous Seven Cities of Cibola. Cortez had the gold placed in wooden boxes and stored in the hold of the ship. He wanted the gold safe and needed to relieve his fighting men of the burden of transporting it.

By the time Ulloa's ship, the *Trinidad,* had reached the mouth of the San Luis Rey River, most of the crew had come down with scurvy. He left three men to guard the ship and took half the gold inland with him. The sailors were too weak to carry the gold more than a few miles, and they buried it near some caves. Ulloa and his men all died without returning to the *Trinidad.*

The three men left on the *Trinidad* waited several weeks for the captain and the rest of the crew to return. Finally, they decided to return to Mexico, figuring the others had died. They could not handle the big ship by themselves, so they left it anchored there and returned to Cortez in a longboat. The *Trinidad* drifted a few miles south, floundered, and went down. Five million in gold went down with it.

One of the three who survived, Pablo Salvador Hernandez, left a diary detailing the history of Ulloa's ill-fated journey. A mass grave was found near Oceanside in 1937. The skulls all showed evidence of scurvy, and pieces of Spanish armor were found in the grave. All this lends support to the story.

One cache of golden coins is known to have been found. Gold coins, thought to come from the *Trinidad,* also have been washed ashore on the beaches near Oceanside. Recently, a treasure-diving crew claimed that they had located the wreckage of the *Trinidad.* What more they have done about salvaging the treasure hasn't been revealed. The five million buried inland remains to be found.

Lost Mission Treasure of Ramona

Folks around the little town of Ramona will tell you that a golden treasure has been hidden near the town since the time when the Indians were the sole residents of the area. The gold is said to have come originally from a mission in Arizona. It was brought here to be shipped to Spain. It was supposed to have been placed in a cave for safekeeping until the ship arrived, and then sealed in by an iron door placed at the entrance of the cave.

Where in the world the Spanish got an iron door no one can say. Rocks and brush were put over the door to hide it. The padres were slain by the Indians, and the treasure cave's location lost forever.

Other lost-treasure stories from this area abound. Warner's Hot Springs and the Vallecito Stage Station each has two lost treasures you might want to research. San Felipe Creek has its treasure tales too. In fact, several artifacts have been found in the creek by people working with metal detectors.

Lost Peg Leg Treasure

One exceptionally intriguing tale is the Lost Pegleg Treasure. The locale of this story is down Banner grade and into the desert. The tale is so well known that a State of California historical marker in the Anza-Borrego State Park honors it. The plaque reads, "Thomas L. Smith, better known as 'Peg Leg' Smith, 1801-1866, was a mountain man, prospector, and spinner of tall tales. Legends regarding his lost mine have grown through the years. Countless people have searched the desert looking for its fabulous wealth. The gold mine possibly could be within a few miles of this monument." They forgot to mention that old Peg Leg was a horsethief and a murderer, too, and that there were two Peg Leg Smiths.

According to the story, Peg Leg had set out for San Bernardino from the Colorado River, near Yuma. As he was crossing the desert, about three days out, a sandstorm hit. He wandered aimlessly for a day, blinded by the driving sand. As the storm died down, he climbed one of three nearby buttes to reorient himself. Resting on top of the butte, he noticed some unusually heavy looking, oddly shaped, black rocks. He stuck a few in his pocket, thinking he would find out later what they were. They were gold, and old Peg Leg spent the rest of his days searching for that same butte without success. The black nuggets traveled for years in the saloons around Southern California, getting Smith free drinks from people who hadn't heard his tale.

Most treasure hunters feel that Peg Leg's nuggets came from a natural placer deposit. I do not. I think this gold came from a pack train that was ambushed by the warlike Yuma Indians. Let me explain.

The Peralta's were a famous family of gold miners from Mexico who are known to have worked mines both in California and Arizona. One of the Peraltas was leading a trainload of ore and nuggets back to Mexico from Northern California, when he decided to spend a few days playing in old Los Angeles. He told the pack train to continue on, and he would catch up with it. He never found it, and when he got to Mexico, he learned that it had never arrived.

The Indians must have driven the pack train off the trail and onto the butte, where it was wiped out. To the Indians, the treasure would have been the mules and horses, not the saddle-bags of gold, which they would have thrown out and left. Over the years, the bags would have decayed and rotted away. What Peg Leg probably found was not a placer deposit but the Peralta gold. The alloys found in the gold from Northern California would cause it to oxidize black in the hot sun of the desert.

For several years, someone has been writing to *Desert* Magazine claiming to have found the Lost Peg Leg. To back up his story, he has sent them several of the black nuggets. I would guess that this treasure probably has been found.

View from Santa Susana Pass.
(Photo courtesy of Security Pacific National Bank)

10 SOUTHLAND TREASURE TALES

Not all the buried treasure stories are set in the gold producing regions. For many years, there were few banks. For even longer, people have distrusted banks. Bandits had no choice. As a result, Mother Earth has received more deposits of wealth than any bank.

Santa Susana Treasure

Coming down out from the mountains and into the San Fernando valley, the first treasure yarn is a romantic one. Its location is in a canyon somewhere along the Santa Susana Pass.

In the early 1850s, a young Mexican cowboy and the daughter of a wealthy Spanish rancher fell in love. The girl's parents rebuffed the young man, telling him he was too poor to marry her. He decided that only by becoming a bandit would he ever be able to acquire enough money to satisfy the girl's parents.

Sadly, for the couple, his first attempt was his last. As the Los Angeles-bound stagecoach reached the summit of Santa Suasana Pass, the lone bandit held it up. He is said to have obtained more than sixty-five thousand dollars from the strongbox and the passengers. One of the passengers, however, had concealed a pistol in his boot, and when the bandit started to ride off with the loot, the passenger drew the pistol and shot him in the back. Although mortally wounded, he was able to remain in the saddle and make his getaway.

When he reached a small canyon running into the pass, he hid there and refreshed himself at a small spring. Now too weak from loss of blood to carry the gold, he buried it near the spring. That night, he made his way to the ranch, where, after telling her what had taken place, he died in the girl's arms. The fate of the gold became known years later when the girl, then an old woman, died, and her diary containing the story was found. No one has found the love loot yet.

Lost Chatsworth Treasure of the Golden Chest

Another treasure tale dealing with lost love is set in the Chatsworth area. An elderly Don from old Mexico, said to be chasing a runaway bride, settled in the little town of Chatsworth around the turn of the century. His health was bad, and he passed away a few years after buying a home there. While alive, he paid for everything with gold coins. He was said to have a chest full of these coins. After he died, a search was made for the chest, but it was never found. A freshly dug hole was discovered under a large oak tree near his house. Folks reasoned that he had moved his gold shortly before his death. Why he had moved the gold no one could guess. When the house was torn down years later, the ground under it was dug up in hopes of finding the coins; but no golden chest was found. That the old Mexican had gold and that it was never found is a fact, say the local old-timers. Since he was so ill, it's unlikely he carried the chest very far.

The Burbank Train Robbery Treasure

In 1893, a Southern Pacific train was derailed at about what is now Roscoe Boulevard in Burbank. The train was wrecked by bandits so they could loot the express car, but one of the train's crew was killed in the wreck. The famous railroad detective Whispering Smith was assigned to the case, and he tracked the bandits to Arizona. They were captured there, but they never revealed where they had buried the money taken in the holdup. It is said that they tried to make a deal for the loot but were turned down. Most people feel it must have been hidden near the derailment spot.

The Buried Pirates Treasure of Northridge

Buried pirate treasure in Northridge? Although it seems highly improbable, a group of shipwrecked Spanish pirates are said to have buried their booty near the corner of Reseda Boulevard and Eddy Avenue in Northridge. There have been several extensive searches made for this treasure in the past. Where the story started or why the Spanish came so far inland to bury their loot is not known. The area is built up today, and the treasure, if it exists, is probably covered by concrete.

Lost Hollywood Gold

Hollywood has long been known as a place of discoveries. In our case, instead of going to Hollywood to be discovered, we are going in hopes of making a discovery.

I heard my first real treasure story as a boy growing up in Hollywood. Sitting against the hills near the Ferndale entrance to Griffith Park is Immaculate Heart of Mary High School and College. The famous bandit Joaquin Murrieta is said to have hidden a large amount of gold, silver, and jewels under a tree behind the grounds.

The Sisters say that from time to time someone will appear at the door with a map he has found in Mexico, showing the location of the treasure. No one has yet been rewarded with the treasure. The worst offenders, the Sisters say, are the treasure hunters who come in the night and leave big holes in the ground. Even when some large earthmoving equipment was used during remodeling, nothing was found. It's there though; just ask any kid in the neighborhood. He'll tell you.

The Cahuenga Pass Treasure

This is the granddaddy of all Los Angeles treasure stories. The treasure has been estimated to be as high as half a million dollars.

The story of the treasure begins in old Mexico during the revolution against the oppressive rule of the foreign Emperor Maximilian. In order to obtain funds to aid his cause, the rebel leader Benito Juarez appealed to the people for help. Gold, silver, and jewels, were collected both in Mexico and California to help finance the revolution.

The funds were entrusted to General Plácido Vega, who arranged to purchase a large amount of arms and ammunition in San Francisco. He selected three of his best officers to handle the transaction. When the riders reached San Bruno in San Mateo county near San Francisco, they stopped for a parley. They must have suspected foul play, as they decided to hide the money and jewels until they could be sure that everything was all right. The treasure then was buried on the side of some nearby hills, and all the evidence of their work hidden. What they didn't know was that a lone sheepherder, Diego Moreno, had observed all that had taken place.

He had been frightened by the sight of such a large number of armed horsemen, because the area had long been known as a place frequented by such bandits as Chavez, Juan Flores, and Tiburcio Vasquez. Moreno thought that these riders were one of those roving outlaw bands. After the Mexicans had departed, he went to the side of the hill to investigate. He dug at the spot where the soldiers had been. When he saw the fortune he had dug up, the sheepherder fled south toward Mexico. In his anxiety to escape with the loot, he abandoned his flock and all his belongings.

When he arrived in Los Angeles, he was almost out of supplies, so he stopped at an inn in the Cahuenga Pass. It was supposedly run by his cousin. While there, he began to worry that some of

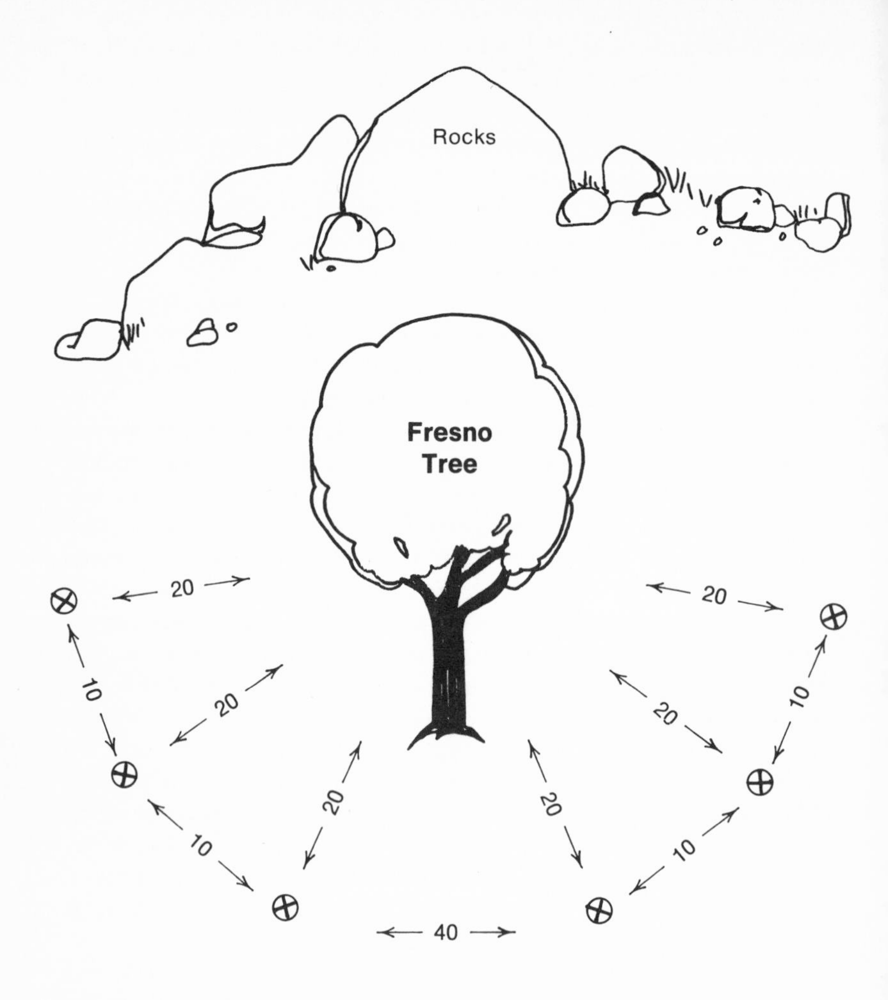

The Cahuenga Pass Treasure Map.
The numbers represent the number of paces
between each buried treasure bag.

the hardcases who stayed at the inn might learn of the wealth he was carrying. He decided to bury it nearby until he had purchased his supplies and was ready to leave. He wrapped the money and jewels in six separate rawhide pouches; then he took the pouches out and buried them. Selecting a Fresno tree on the east side of the pass above the inn as a location marker, Moreno placed the pouches in six holes at equal distances from the tree. The holes were arranged in the shape of a crescent.

The original map to the site was shown to me years ago by an old man who swore it was authentic. He said that one day we would go out and recover the cache together; but he passed away a short while later, and the map was never found.

After he had buried the money and the jewels, Moreno returned to his room at the inn. That night, while he slept, he was attacked and severely wounded. Whoever assaulted him must have suspected that he was hiding something, because the room was ransacked, and the packs torn open. The critically wounded Moreno was found the next morning more dead than alive. Unable to care for him properly at the inn, his cousin took him to the home of some friends in Los Angeles named Martinez. The Martinez family did all they could for the sheepherder, but it was to no avail as he only grew weaker during the next few weeks. When he realized he was dying, Moreno decided to reward the family for their kindness. Calling the head of the family, Jesus Martinez, to his bedside, Moreno told him of the great fortune he had hidden, and where he had buried it. If he could not recover, the fortune was theirs, Moreno said. The elder Martinez drew a crude map based on the directions Moreno had given him. A short time later, the sheepherder died.

Martinez, with a boy named Joseph Corea, who later became a policeman in Los Angeles, went to dig up the treasure. With the map to guide them, they located the tree with little difficulty. The excitement was too much for the elderly man, who began to complain of chest pains and asked Corea to wait while he sat down to rest. He never got up again. His heart had failed him. The boy became terrified when the man died, and he fled. From that day forward, Corea refused to return to the site, believing the treasure to be cursed.

Nothing more was heard of the treasure until 1885. It was in that year that a Basque sheepherder had a turn of luck that would change his life forever. For several years previous to 1885, the sheepherder, whose name was said to be Correo (this name is also given to Moreno's cousin in some accounts), had roamed the hills surrounding the Pass with his flock. Each day he would stop and eat his lunch under the same tree. The spring of 1885 was one of unusually heavy rains. Shortly after one of the hardest of the storms, Correo made his discovery. He was eating his lunch when he noticed his dog pulling something from a hole he had been digging. His curiosity aroused, Correo walked over to see what the dog was tugging at. Just as he got there, the dog managed to pull the object from the ground. As it came loose, it unrolled at the sheepherder's feet—gold, silver coins, diamonds, rubies, emeralds, and jewelry of all sizes and shapes gleamed and sparkled before his eyes. The dog, of course, had unearthed one of the six rawhide rolls, worth nearly one hundred thousand dollars, buried by Moreno twenty years before.

The sheepherder took the treasure and returned to Spain a rich man. He did not know the story of hidden treasure, and so he never realized that there were five more rolls buried under the same tree.

Another find was made in 1943, when a treasure hunter with a metal detector turned up a small amount of gold and silver coins from this area.

Several years ago, the city gave two treasure hunters permission to search the Hollywood Bowl parking lot with metal detectors. They were to share with the city any treasure found, but they were unable to locate anything and, after several days, they gave up their hunt. Most treasure hunters feel that the area around the Bowl is the most likely spot for the treasure to have been hidden.

The Lost Hollywood Gold Mine

Somewhere in the Hollywood Hills there is a lost gold mine. The mine was owned by a city official around the turn of the century. It was somewhere near the end of the old trollycar line.

Many old adobe buildings have treasure stories attached to them. (Photo courtesy of Title Insurance and Trust Company)

Big Tujunga Canyon was the site of several rich strikes.
(Photo courtesy of Security Pacific National Bank)

11 KNOWN GOLD-BEARING AREAS

There are several other areas you should prospect in your search for gold. All of the following regions have produced gold in various amounts.

Tujunga

There are two Tujunga Canyons, Big Tujunga, and Little Tujunga. To reach Big Tujunga Canyon, go north on the Golden State Freeway (Interstate 5), to the Sunland Boulevard turnoff. Go right to Foothill Boulevard. Turn right on Foothill Boulevard and continue on to Mt. Gleason Avenue. Go left on Mt. Gleason. Mt. Gleason turns into Big Tujunga Canyon Road. Gold Creek runs off Big Tujunga Canyon Road. Several thousand dollars in placer gold have been taken out of the creek gravels here.

The Little Nugget Claim has been worked on and off from 1918 to 1947. (There is even a fossil bed on a bench above the creek that we found one day.) Some people include the Mill Creek deposits and the Monte Cristo Group with the Big Tujunga region, but I put them with the Acton District because that is how the State Department of Mines and Geology lists them. From a geographical point of view, however, they are closer to Big Tujunga.

There are both lode and placer deposits in this Canyon and the smaller canyons running into it. One of the most productive areas has been the Dutch Louie Placers in Pacoima Canyon. Some silver and titanium has been mined along with the gold in the Canyon.

Mount Gleason

Higher up in the San Gabriel Mountains is Mount Gleason. The first mine here was discovered by a man named Gleason who went up the slope to harvest lumber for the mines at Acton. The most productive mines were the Los Padre and Mount Gleason. Relatively small amounts of gold were mined in the district.

Mount Baldy

This region is just west of Mt. San Antonio (Old Baldy Peak). The San Gabriel River placers are sometimes included in this district. The mines, set high up on the mountain slopes, took a lot of guts to locate and to mine. They were discovered by prospectors seeking the mother lode that feeds the placers of the east fork of the San Gabriel River. Active mining was done here as late as 1941, although the principal mining was done from 1903 to 1908. The lode gold mines have produced over fifty thousand ounces of gold, the largest producer being the Big Horn Mine. The gold occurs in quartz veins in schist. The surface ore has been richest, producing some beautiful jewelry rock.

Neenach

Neenach is a small district in the foothills facing the Antelope Valley. Take the Golden State Freeway (Interstate 5) north to Highway 138. Go east on Highway 138 to County Road N2. The mines were in the hills on your right. (Be careful here, as there are several open shafts and you could fall in.) Gold was first found here in 1899. The greatest era of production was from 1935 to 1938, when over two hundred thousand dollars in gold were taken out. Most of that came from the Rogers-Gentry group of mines. The gold occurs in quartz veins with pyrites.

Big Rock Creek

Prospectors have been having some success recently in Big Rock Creek. This is one of the most beautiful regions in the Angeles National Forest, where the trees remind me of those in the High Sierras. Take the Golden State Freeway (Interstate 5) to the Antelope Valley Freeway (Highway 14). Go east on the Antelope Valley Freeway to Highway 138. Go right on Highway 138 to the town of Little Rock. Take the Valyermo Road out of Little Rock to Big Rock Creek. There are several nice campgrounds if you want to spend some time here.

Santa Ana River

For years I was told that there was no gold in the Santa Ana River. Then one day an old-timer said, "Don't you believe it," and he told me exactly where there *is* gold in the Santa Ana River. Now I'll tell you. I have found bigger pieces of gold here than in the much more publicized Holcomb Valley area.

From Los Angeles take the San Bernardino Freeway (Interstate 10) to Redlands. Take Highway 38 out of Redlands, all the way up into the San Bernardino Mountains. Before you come to Big Bear, Highway 38 crosses the Santa Ana River. There is a dirt road on your left, called River Road. This road is near the source of the river. Work anywhere along the river for a mile or two, and you will find gold. There is a nice campground here as well.

Prospecting Tip

Keep in mind that many geologists consider the Peninsular Range of mountains to be the southern Sierra Nevadas. (The Peninsulars are the mountains that run south beginning in the Santa Ana Canyon.) This is because they are the same age and composition as the northern Sierras. Try your luck in these mountains, and maybe you will discover a new mother lode.

Burros packed for trip to the mines.
(Photo courtesy of Title Insurance and Trust Company)

Early mining methods including the use of (from upper left) a tunnel, rocker, pan, sluice, shaft, long tom, and coyote hole. (Photo courtesy of Security Pacific National Bank)

HOW TO FIND GOLD

12 INTRODUCTION

Now that you know where the gold is, let's find out how you can get some into your poke. Interestingly enough, terms that had their beginnings in the Gold Rush of 1849 are still used by today's prospector.

Prospecting simply means that you are testing the prospects of each area you visit by sampling until you find the spot where you can get the most gold. You sample by taking a pan full of dirt from a spot that looks promising, and then you pan it down to see what you get. You may want to bring home samples of rocks to check more carefully. If you do, be sure to keep track of where you found them. Just take a roll of white tape along, write the location on a piece, and stick it on the rock. There must be a thousand stories of lost bonanzas in which someone picked up a pretty rock, stuck it in his pocket, and found out later that it was fabulously rich ore. Of course, the person could never find his way back to the particular spot the rock came from.

Over the years I've picked up a few techniques that make amateur prospecting easier. A big washtub at home to pan out any dirt you bring home is a good idea. Be sure that you always bring a warm jacket with you—the Southern California nights can get cold even in the summer, as you probably know. A pair of waterproof boots are an absolute necessity. You may want to wear gloves to protect your hands as well. Heavy rubber gloves are excellent in the winter when the water is *really* cold in the creeks and streams. Be sure that you bring along a canteen or a jug of water when you are out. Prospecting can be hard work, and you can really get dried out.

There are two types of gold mining, lode and placer. Lode, or hardrock mining, takes place at the gold's source, mostly in quartz veins. When an outcropping of gold-bearing rock is found, it is mined by following the vein, either by a tunnel or a shaft, depending on the dip of the vein. An easy way to tell the difference between a tunnel and a shaft is that you walk into the former and fall into the latter. Seriously, a tunnel goes into the side of a mountain and a shaft goes down into the ground.

More than likely you will be placer mining, or placering, in the beginning. There are several types of placers including bajada, eluvial, residual, marine, and stream placers. Of all the different types, the ordinary stream placer is by far the most important. The majority of the gold mined in California has come from placer deposits in either present or ancient stream gravels. These deposits occur where a stream flows over an exposed vein or where the gold is washed down into the streambed from the surrounding countryside. Occasionally, several low-grade veins create a rich placer deposit.

When prospecting, always keep in mind that the gold is nearly six times as heavy as most gravels. The gold, therefore, will work its way down to or near the bedrock. It also will settle quickly wherever the stream course is slowed down. The best places to look for placer gold are on the downstream side of obstructions, such as boulders. Gold also will drop on the inside of curves made by the stream, and on the bars where the stream widens. Look carefully in the roots of trees alongside a riverbed. Even the roots of small plants and moss along the stream's edge will trap particles of gold. The richest spots will be in cracks or crevices in the streambed. (Up north in the Mother Lode, the gold was found everywhere, even in valleys and on hills where water must have flowed millions of years ago.)

Another good source of rich placer deposits are gravel benches sitting high on the canyon walls of present streams. These benches were once the bed of the stream now running far below them. You can dig right on the ancient bedrock if you can reach these benches.

Now, about those bright, shiny, glittering flakes you see in the streambeds and on the benches. Well, those bright flakes are material proof that all that glitters is not gold. Called fool's gold, these particles are technically known as iron pyrite. Once you find your first piece of gold, you'll never be fooled again. Fool's gold is much lighter in weight than gold, will float in your pan, and normally will be on top of the black sand. A simple test can be used to find out whether you have gold or iron pyrite. Take a knife blade and try to cut the particle in question; if it shatters, it's fool's gold.

I remember one time I was working in one of the local creeks when a group of young people came walking down from upstream all excited. I could see that they had two milk bottles full of black sand and pyrite. They were discussing how they were going to spend their new-found fortune. When they saw me, they stopped and watched me shovel dirt into the two sluice boxes I had working. Hiding the bottles behind their backs, they asked

me what I was doing.
"Prospecting," I said.
"For what?" one asked.
"Gold," I answered.
"Well, what's that?" he inquired, pointing to the glittering fool's gold in the creekbed.
"Pyrite, fool's gold," I told him.
It was a sad but wiser group that departed. With the help of this book you won't ever go home without a little color for your efforts.

If you are still not sure if your colors are real gold or fool's gold, take your sample to a prospectors' supply store or a jeweler. Experts at either place will tell you what you have.

Youngster panning a creek. (Photo by James Klein)

Old prospector with his burro.
(Photo courtesy of Title Insurance and Trust Company)

13 EQUIPMENT

The Pan

The pan is still the basic item used by both the beginner and the old pro. During the rush to the Mother Lode, almost anything that was available was used as a gold pan. Many old-timers cooked and panned out of the same skillet.

The Mexicans were probably the first to bring a vessel to the gold fields for the express purpose of separating precious metals from the sand and gravels of streambeds. The Mexican pan was called a *batea*, and it was carved from wood. It was fifteen to sixteen inches wide, six to eight inches deep, and fairly heavy. An interesting story regarding the *batea* is told about James Marshall, the discoverer of gold at Sutter's Mill. This discovery started the California Gold Rush, of course.

It seems that one of the workers at the mill, who was from Mexico, had told Marshall that if he had a *batea*, they could recover gold from the gravels of the river for themselves. Marshall thought that the *batea* was some sort of complicated device, so he put the man off. Whether one *batea* would have helped him, we'll never know, but Marshall saw little of the millions in gold that resulted from his discovery. He died a penniless and bitter man.

Today's pans are made of steel, copper, or plastic. For the beginner, I would recommend the plastic pan for several reasons. One reason is that it is molded with a set of cheater riffles in the pan. These ridges keep a new panner from losing a lot of his gold by acting as traps for the gold. The pans are black, so the color contrast with the gold makes the metal easier to see when separating it from the concentrates. Steel pans, on the other hand, must be burned black when they are new. Another good thing about the plastic pan is that it will not rust or become corroded if wet black sand is left in it. The pros use copper pans when using mercury to recover fine gold.

Three sizes of pans are made today. The small sampling pan is about six to eight inches wide; the medium pan, which most beginners find easiest to handle, is twelve to fourteen inches wide; the largest or professional size pan is sixteen to eighteen inches wide. Stay away from the big one in the beginning because it is awkward to carry around and collects too much material for easy panning. It also gets pretty darn heavy. Most hardware stores carry the steel pans. The plastic and copper pans can be purchased at prospectors' supply stores, rockhound shops, and some sporting goods stores.

The Shovel

You'll need a good shovel to dig up the dirt to put into your pan. Take your choice, long handled or short. The long handled shovel is easier on the back; the short handled one is easier to carry. You also should have a small hand garden trowel, for the hard-to-get-at spots.

Tweezers

You will need a pair of the pointed jeweler's tweezers to pick the tiny flecks of gold from your pan.

Small Bottle

Find a small bottle or vial of clear glass or plastic to hold your colors.

Magnifying Glass

Any small magnifying glass that you can carry in your pocket will do. For home study, the magnifying glass attached to a flashlight is the best.

Knife

A plain hunting knife will do. Use this for digging, prying, and scraping.

Brush

A small paint brush can be used for cleaning out cracks and dusting crevices which are hiding gold, hopefully.

Bucket

Any regular household plastic bucket will suffice. The size will depend on how much you will want to carry.

A Prospector's Hammer

You may want to invest in a prospector's hammer, which is used to split rocks to see if they contain any valuable minerals. You can also use the hammer end to pry with. It is pointed on one end and flat-headed on the other. Buy a good one, otherwise the point on the cheaper ones will flatten out after very little use.

These are the basic tools you should have to start prospecting. As you go on in your golden quest, you will want to acquire more sophisticated equipment such as a sluice box, dry washer, dredge, concentrator, or metal detector.

The Sluice Box

Once you have mastered the pan, the next piece of equipment you will want is a sluice box, the most widely used placer mining item next to the pan. A sluice box will enable you to wash at least five to ten times as much dirt as you can pan in one day. The sluice box is a fairly simple device, and it can be made at home or bought ready-made. The sluice uses the flow of the stream, creek, or river water to separate the gold from other materials and to trap it behind a series of obstructions (or riffles) in the box.

The sluice has been used by goldseekers since before the Gold Rush. The first ones were heavy and difficult to manage, and so are some modern ones. I have found quite a few wooden, homemade sluice boxes abandoned, because they were heavy and ineffective. Because of this, sluices were built right on the spot in the 1849 Gold Rush. Some of the sluices built then were as long as a quarter of a mile. These spectacularly long sluices were called "Long Toms." Several men were required to work them; some would bring the mineral rich soil to the sluice, while others would shovel it into the box and keep it moving.

The shape of the sluice remains basically the same today as it was then; however, it is now made from lightweight metals, so it is more portable. The riffle designs are also more effective. A sluice box, in reality, is simply a trough about ten to twelve inches wide, with six inch high sides, and obstructions called riffles about one-half to three-quarters of an inch thick, fitted every six inches along the bottom. The sluice box is placed in a fast-flowing part of the stream with at least four inches of water. The foot of the sluice is always lower than its head.

Material is shoveled into the head of the sluice box; the force of the water pushes the material down the sluice and over the riffles, where the black sands and gold are trapped. The water should be moving fast enough to break up the material in about thirty seconds, but not so fast that the gold is carried off. The sluice box is cleaned up once or twice a day by washing the concentrates into a pan, panning out the gold that has been recovered, and placing it in containers.

Keene Electrostatic Concentrator which can be used where no water is available. (Photo courtesy of Keene Engineering)

The Dry Washer

With the lack of water in so many streams in Southern California most of the year, a dry washer becomes a necessity if you are really serious about prospecting. Lack of water has been a major reason for the curtailment of gold production in this region. The dry washer works on the same principal as the sluice box, using forced air instead of water. There are many different types of dry washers at prospectors' supply houses. You can also build one yourself.

The Keene Electrostatic Concentrator

The Keene Electrostatic Concentrator is the best device for dry washing that I have come across. This dry washer uses a high-static fan to force air up into a plastic trough, where it gains an electric charge from the plastic. The air then moves under pressure through a special artificial fabric where the charge is increased. Material is shoveled into the concentrator through a screen which automatically classifies the material, letting only gravel less than half an inch wide into the concentrator. (Don't worry about losing any nugget bigger that that. If one is on the screen, you will see it, believe me.) The material then works its way down to the plastic tray. Gold is non-magnetic, but it has an affinity for an electrostatic charge, and it will stick to the special cloth. The concentrator also works like a regular dry washer and traps the gold behind riffles.

The Dredge

In the past, using a dredge was a complicated procedure. It has been used in California since the 1880s, but today's dredges are lightweight and portable, so they can be worked by one man. The surface dredge is most commonly used. It consists of a portable sluice box with a vacuum hose attached for dredging gravels from under the surface of streams and rivers. The gold-bearing gravels are vacuumed up by the hose and pumped to the sluice box where the gold is trapped by the riffles.

Several sizes of dredges are on the market. Portable ones run one and a half to six inches in hose diameter. The bigger they are, the heavier they are. The nicest thing about a dredge is that it sure beats digging.

The Metal Detector

A booklet called *Finding Lost Treasure,* put out by Detectron (a manufacturer of metal detectors) states, "The detector has never

Small dredge used for vacuuming gold particles from the sides and bottom of a stream. (Photo courtesy of Keene Engineering)

Larger and more powerful dredge used for underwater prospecting. (Photo courtesy of Keene Engineering)

been built which can positively select only gold." Due to this fact, I would not recommend using a metal detector for gold prospecting. However, a metal detector can be used to locate mineralized areas and heavy concentrations of black sand.

Some people claim to have mastered the art of nugget shooting, which is finding good-sized nuggets with the metal detector. I use my detector mainly for hunting artifacts. A metal detector can be fun, and some users actually pay for their detectors with the coins they find.

This is a general list of prospecting equipment you would be interested in owning. (Keene Engineering will send you a free equipment catalog if you write them at 9330 Corbin Avenue, Northridge, California, 91324.)

The pan with cheater riffles moulded into it.
(Photo courtesy of Keene Engineering)

14 HOW TO PAN

Panning is relatively simple, but it takes a little practice. Learn it in the following steps:

1. Find a spot in the stream where the water is sufficiently deep to submerge your pan completely, and it is flowing fast enough to carry off the muddied water.
2. Fill your pan one-half to three-fourths full of the dirt or gravel you wish to wash. Dip the pan into the stream and fill it with water. Holding the pan with one hand, use the other hand to break up any clods that may be in your material.
3. With a gentle circular motion, make the gravel swirl around the pan. This will cause the dust and clay to come to the top.
4. Submerge the pan again and continue the circular motion allowing the lighter material to float away.
5. When the water has cleared, pick out the largest pebbles and throw them away.
6. Keeping the pan just under the water, tip it slightly away from you. Begin swirling the water around the pan and with a slight, forward, tossing motion, carefully forcing the lighter materials out of the pan. When using the plastic pan, be sure that the cheater riffles are toward the front, so that the gravel is passing over them as you toss the gravel out.
7. From time to time, level the pan, and shake it back and forth. This will cause the lighter materials to come to the top and the gold to sink to the bottom.
8. Repeat the above process until only the heavier material remains. This will usually be black sand with the gold underneath. These are your concentrates or black sand.
9. Take your pan out of the stream and pour off most of the water, leaving enough to cover the black sand by about half an inch. Shake everything to the front of the pan by tilting it forward. Now tilt the pan toward you and start the swirling again until the black sands are removed and only the gold remains. Use your tweezers to pick out the flakes, and put them in your specimen bottle for safekeeping.

When using the plastic pan you can use a magnet to help you in the final processing. Place the magnet on the underside of the pan and tilt the pan slightly while moving the magnet in a small, circular motion. This will separate the magnetic black sand from the gold.

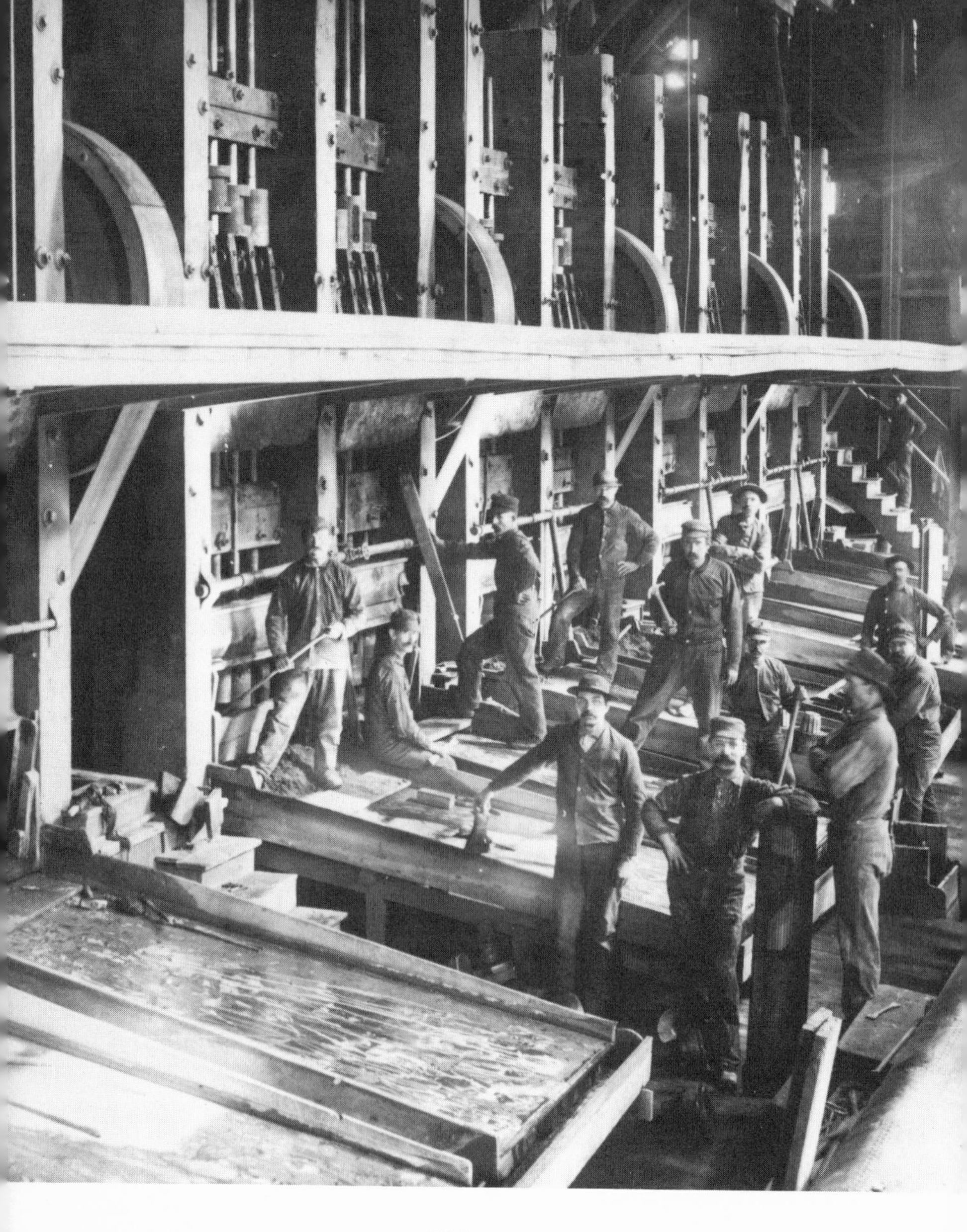

Old stamp mill.
(Photo courtesy of Title Insurance and Trust Company)

15 HOW TO STAKE, ASSAY, AND REFINE

If, by chance, the Great Miner in the Sky should smile on you, and you locate a rich deposit of gold, you will need to know how to claim it for your own. Remember that the first men involved in writing California law were either miners or men connected with the Gold Rush in some way. Men like Hearst, Armour, and Stanford, whom we associate with newspapers, meat packing, and universities, began their fortunes in the gold fields. It was these same men who helped draft California's laws.

The state was dependent on the gold-mining industry, and its land laws still favor the miner. If you own your home, you probably are aware that you do not get mineral rights with your title. Mineral rights have to be claimed and proved if contested. Mineral rights belong to the locater.

If you locate a deposit you feel is worth developing, there are some basic things you can do to protect your discovery. First, as you know, there are two types of gold mines: placer and lode. Each has its own set of rules and regulations.

A placer claim may consist of as much as twenty acres for each person signing the location notice. Placer gold, again, is free gold found in gravels and alluvial deposits. You must place a Notice of Location on a post, tree, rock in place, or on a rock monument you build, showing the name of the claim, who is locating it, the date, and the amount of area claimed. You must mark the boundaries of your claim so that it may be easily traced. You must also identify the location of your claim by reference to some local landmark or natural object (such as a stream or rock formation) or a permanent monument. Finally, you must file a copy of the Location Notice with the local county recorder within ninety days of the date of location.

A lode claim uses a different procedure. Lode gold, again, is gold in a vein, which must be mined and separated from its mother rock. The claim may consist of as much as 1500 feet along the sidelines of the vein, and 300 feet on each side of the middle of the vein. Within ninety days of the location of any lode-mining claim, you must place a post or stone monument at each corner of the claim. The posts must be at least four inches in diameter, and the stone monuments must be at least eighteen inches high. A Notice of Location form must be posted on the claim, and a lode Location Notice filed with the county recorder within ninety days.

NOTICE OF LOCATION

(SECTS. 2301 TO 2324, PUBLIC RESOURCES CODE)

TO WHOM IT MAY CONCERN: Please take notice that:

1. The name of this claim is the ______________________________ Lode Mining Claim. Said claim is situated in the ______________________________ Mining District, County of ____________________, State of ____________________ Located this ______________ day of ____________________, 19______, when discovery was made and copy of the original notice posted.

2. That the undersigned locators are citizens of the United States or have declared their intention to become such, and have discovered mineral bearing rock in place thereon and do locate and claim same for mining purposes.

3. That the number of linear feet in length along the course of the vein each way from the point of discovery whereon we have erected a monument, is ______________erly______________feet; ______________erly ______________feet; that the total length does not exceed 1500 feet. That the width on the______________erly side is______________feet; that the width on the______________erly side is______________feet; that the total width does not exceed 600 feet, that the end lines are parallel; that the general course of the vein or lode as near as may be, is in a______________erly and ______________erly direction; that the boundaries of this claim may be readily traced, and are defined as follows, to wit:

COMMENCING at ______________ in a ______________erly direction from the discovery monument, ______________ feet therefrom, ______________________________

____

____

____

____

____

Township:______________________________Range:______________________________

4. That this claim can be identified by its proximity to the following natural object or permanent monument, to-wit: ______________________________

____

____

This notice is intended as an original notice of location and is posted on this claim and a true copy thereof is to be recorded with the County Recorder of this County.______________________________

____

____

____

That all the dips, variations, spurs, angles and all veins, ledges, or deposits within the lines of this claim, together with all water and timber and any other rights appurtenant, allowed by the law of this State or of the United States are hereby claimed.

LOCATORS

______________________________ ______________________________

______________________________ ______________________________

______________________________ ______________________________

______________________________ ______________________________

WITNESSES

______________________________ ______________________________

______________________________ ______________________________

The exterior boundaries of a Lode Claim cannot be limited by any local mining regulation to less than 25x1500 feet, measuring from the center of vein on either side.

PUB. RES. CODE 2313, within ninety days after the posting of this notice of location upon a lode mining claim, placer claim, tunnel right or location, or mill site claim or location, the locator shall record a true copy of the notice together with a statement of the markings of the boundaries as required in this chapter, and of the performance of the required discovery work, in the office of the County Recorder of the County in which such claim is situated.

LOCATION NOTICE—LODE—WOLCOTTS FORM 1133—REVISED 2-68

Sample of a lode gold Notice of Location.
(Courtesy of Wolcotts)

This section is not intended to be a complete thesis on all the rules and regulations regarding mining claims, but it will help to protect you in the initial phases should you make a rich discovery. For more details, I recommend obtaining a copy of the *Legal Guide for California Prospectors and Miners,* Special Publication 40, put out by the State of California Department of Mines and Geology.

Assaying and Refining

If you think you have found a good gold spot, you'll want to find out its real value. (Incidentally, a weekend prospector should not tackle amalgamation casually. Mercury is a very dangerous material, and it should be used only by those well trained in handling it.) Take your samples to an assayer. He will give you a full report on your sample. Who knows, you might have also located a silver mine as well!

If your assay shows a good percentage of gold, you have several options. You can contact a mineral resource company to help you develop the lode, or you can develop the mine yourself.

In the latter case, you'll ship the material to a smelter to be refined and returned to you. The refinery may buy it directly from you as well.

If you're successful at placering, you can sell your gold directly to a refinery, jeweler, dentist, assayer, or prospectors' supply store.

Check the paper for the current price of gold before approaching a possible buyer. Your unrefined gold will sell for less, naturally, than the refined gold price quoted in the paper; however, the daily market price will give you a ball-park figure to start from.

The following refineries are close by:

Handy and Harman
4140 Gibson Road
El Monte, Ca. 91731

Martin Metals, Inc.
1321 Wilson Street
Los Angeles, Ca. 90021

PGP Industries, Inc.
13429 Alondra Boulevard
Santa Fe Springs, Ca. 90670

Industrial Silver Company
1120 North Citrus Avenue
Los Angeles, Ca. 90038

Western Alloy Refining Company
366 East 58th Street
Los Angeles, Ca. 90011

Good luck, good hunting, and I'll see you out there!

NOTICE OF LOCATION

PLACER CLAIM

NOTICE IS HEREBY GIVEN: *That the undersigned citizens of the United States, over the age of twenty-one years, in compliance with the requirements of Chapter VI, Title 32, of the revised Statutes of the United States and the local customs, laws and regulations, ha____ this day located and claimed the following described Placer Mining grounds, viz:*__________

____

together with all water and timber appurtenant, allowed by law, are hereby claimed.

THIS CLAIM *consisting of* ____________________ *acres, or*____________________

*number of feet claimed, shall be known as the*______________________________

*District, County of*____________________, *State of*____________________,

Section ____________________, *Township* ____________________,

Range ____________________, *Meridian* ____________________.

THIS CLAIM *to be identified by its proximity to the following natural object or permanent monument, to-wit:* ______________________________

LOCATED THIS __________ *day of* ____________________, 19____.

The date of discovery and posting of this notice is the _____ *day of* ________________

19_____.

LOCATORS:

____________________ ____________________

WITNESSES

____________________ ____________________

The exterior boundaries of a Placer Claim cannot be limited by any local mining regulation to less than 25x1500 feet, measuring from the center of vein on either side.

PUB. RES. CODE 2313, within ninety days after the posting of this notice of location upon a lode mining claim, placer claim, tunnel right or location, or mill site claim or location, the locator shall record a true copy of the notice together with a statement of the markings of the boundaries as required in this chapter, and of the performance of the required discovery work, in the office of the County Recorder of the County in which such claim is situated.

LOCATION NOTICE—PLACER—WOLCOTTS FORM 1120 - REV. 1941

Sample of a placer gold Notice of Location. (Courtesy of Wolcotts)

GLOSSARY

AURIFEROUS. Containing gold or gold bearing.

ARRASTRE. A circle of stones where ore was crushed during the early days of gold mining; a primitive but effective method of separating gold from quartz.

ASSAY. To evaluate the quantity and quality of the minerals in an ore.

BAR. A name given to the sandbars and rock and gravel bars found in rivers, primarily when they are gold-bearing.

DIGGINGS. A claim or place being worked.

DIORITE. A granular, crystalline, igneous rock in which gold sometimes occurs.

DREDGING. A method of vacuuming gold-bearing gravels from river or stream bottoms.

DRY WASHER. A machine which separates gold from gravels by the flow of forced air.

GLORY HOLE. A small but very rich deposit of gold ore.

GRAVEL BENCHES. Gravel deposits left on canyon walls through stream erosion.

HARDROCK MINING. Another term for lode mining.

HEADFRAME. The support structure located at the entrance of a mine over a shaft. Used for hoisting.

HYDRAULIC MINING. A very destructive and now outlawed form of gold mining used in the Gold Rush. Giant hoses were used to force great streams of water onto canyon walls containing gold-bearing gravels. The walls were washed away into sluice boxes, where the gold was then picked out.

IRON PYRITE. A common mineral consisting of iron disulfide which has a pale, brass-yellow color and a brilliant metallic luster. Also called fool's gold.

LODE. A vein of gold mined either through a tunnel or a shaft.

MATRIX. The material in which the gold is found.

PLACER. Free-occurring gold which is usually found in stream and river gravels.

POCKET. A rich deposit of gold occurring in a vein or in gravels.

POKE. A leather pouch used by old-time miners to hold their gold.

QUARTZ. A common mineral, consisting of silicon dioxide, that often contains gold or silver.

RETORT. A device used to separate gold from mercury.

RICH FLOAT. Gold-bearing rocks worked loose from a lode.

ROCKER. A device used by the early miners during the Gold Rush. This was a sluice box mounted on rockers with a hopper on the top to classify the material. Gravels were shoveled into the hopper; then water was poured on top, washing the gold-bearing material down over the riffles while the hopper was rocked. The rocking helped the gold to settle.

SCHIST. A crystalline rock which is easily split apart.

SLUICE BOX. A trough with obstructions to trap gold used in continuously moving water.

STAMP MILL. A machine used to crush ore.

SULFIDE. A compound of sulfur and any other metal.

TAILINGS. The material thrown out when ore is processed. The tailings from the early mines, where the miners were sometimes very careless, have produced significant amounts of gold and other valuable minerals.

TERRACE DEPOSITS. Gravel benches high on canyon walls.

WIRE GOLD. Gold thinly laced through rock.

James Klein was born in Beechgrove, Indiana, in 1932, and arrived in California at the age of 14. After attending Los Angeles City College and U.C.L.A., he went to work at the old *Daily News*. He later moved to the *Los Angeles Mirror-News*, and then on to the *Los Angeles Herald-Examiner*. For the next five years, he was a sales manager of an automotive battery company.

During his years as a newspaperman, he began doing small parts in motion pictures and television. He now devotes his full time to acting and prospecting. Introduced to gold prospecting ten years ago by a friend, he caught gold fever almost at once. A partner in K.&M. Mining Explorations Company, which is now developing three gold mining claims, he spends most of his time in the field looking for new deposits. When not in the field, he does television commercials or guest appearances on various television shows.

Today the world economy has gathered tremendous momentum—it is both huge in size and growing fast. But what is its direction?

I am seated in my study as I write this, looking at a stack of books about two feet high. They share a common theme, and it is not a happy one to contemplate. We can see this theme immediately in their titles.[21]

By a conservative jurist: Richard A. Posner, *Catastrophe: Risk and Response*

By the president of the Royal Society in the United Kingdom: Martin Rees, *Our Final Hour: How Terror, Error and Environmental Disaster Threaten Humankind's Future*

By a leading American scholar: Jared Diamond, *Collapse: How Societies Choose to Fail or Succeed*

By a British scientist: James Lovelock, *The Revenge of Gaia: Why the Earth Is Fighting Back and How We Can Still Save Humanity*

By an American expert: James Howard Kunstler, *The Long Emergency: Surviving the End of Oil, Climate Change, and Other Converging Catastrophes of the Twenty-first Century*

By a U.S. expert on conflict: Michael T. Klare, *Resource Wars: The New Landscape of Global Conflict*

By an Australian diplomat and historian: Colin Mason, *The 2030 Spike: The Countdown to Global Catastrophe*

That is but a sample of the "collapse" books now on the market. Each of these authors sees the world on a path to some type of collapse, catastrophe, or breakdown, and they each see climate change and other environmental crises as leading ingredients of a devil's brew that also includes such stresses as population pressures, peak oil and other energy supply problems, economic and political instabilities, terrorism, nuclear proliferation, the risks of various twenty-first-century technologies, and similar threats. Some think a bright future is still possible if we change our ways in time; others see a new dark ages as the likely outcome. For Sir Martin Rees, "the odds are no better than fifty-fifty that our present civilization on earth will survive to the end of the present century."[22] Personally, I cannot imagine that the risks are so great, but Rees is a thoughtful individual. In any case, it would be foolish to dismiss these authors. They provide a stark warning of what could happen.

The escalating processes of climate disruption, biotic impoverishment, and toxification that continue despite decades of warnings and earnest effort constitute a severe indictment, but an indictment of what exactly? If we want to reverse today's destructive trends, forestall further and greater losses, and leave a bountiful world for our children and grandchildren, we must return to fundamentals and seek to understand both the underlying forces driving such destructive trends and the economic and political system that gives these forces free rein. Then we can ask what can be done to change the system.

The underlying drivers of today's environmental deterioration have been clearly identified. They range from immediate forces like the enormous growth in human population and the dominant technologies deployed in the economy to deeper ones like the values that shape our behavior and determine what we consider important in life. Most basically, we know that environmental deterioration is driven by the economic activity of human beings. About half of today's world population lives in abject poverty or close to it, with per capita incomes of less than two dollars a day. The struggle of the poor to survive creates a range of environmental impacts where the poor themselves are often the primary victims—for example, the deterioration of arid and semiarid lands due to the press of increasing numbers of people who have no other option.

But the much larger and more threatening impacts stem from the economic activity of those of us participating in the modern, increasingly prosperous world economy. This activity is consuming vast quantities of resources from the environment and returning to the environment vast quantities of waste products. The damages are already huge and are on a path to be ruinous in the future. So, a fundamental question facing societies today—perhaps *the* fundamental question—is how can the operating instructions for the modern world economy be changed so that economic activity both protects and restores the natural world?

With increasingly few exceptions, modern capitalism is the operating system of the world economy. I use "modern capitalism" here in a broad sense as an actual, existing system of political economy, not as an idealized model. Capitalism as we know it today encompasses the core economic concept of private employers hiring workers to produce products and services that the employers own and then sell with the intention of making a profit. But it also includes competitive markets, the price mechanism, the modern corporation as its principal institution, the consumer society and the materialistic values that sustain it, and the administrative state actively promoting economic strength and growth for a variety of reasons.

Inherent in the dynamics of capitalism is a powerful drive to earn profits, invest them, innovate, and thus grow the economy, typically at exponential rates, with the result that the capitalist era has in fact been characterized by a remarkable exponential expansion of the world economy. The capitalist operating system, whatever its shortcomings, is very good at generating growth.

These features of capitalism, as they are constituted today, work together to produce an economic and political reality that is highly destructive of the environment. An unquestioning society-wide commitment to economic growth at almost any cost; enormous investment in technologies designed with little regard for the environment; powerful corporate interests whose overriding objective is to grow by generating profit, including profit from avoiding the environmental costs they create; markets that systematically fail to recognize environmental costs unless corrected by government; government that is subservient to corporate interests and the growth imperative; rampant consumerism spurred by a worshipping of novelty and by sophisticated advertising; economic activity so large in scale that its impacts alter the fundamental biophysical operations of the planet—all combine to deliver an ever-growing world economy that is undermining the planet's ability to sustain life.

The fundamental question thus becomes one of transforming capitalism as we know it: Can it be done? If so, how? And if not, what then? It is to these questions that this book is addressed. The larger part of the book proposes a variety of prescriptions to take economy and environment off collision course. Many of these prescriptions range beyond the traditional environmental agenda.

In Part I of the book, Chapters 1–3, I lay the foundation by elaborating the fundamental challenge just described. Among the key conclusions, summarized here with some oversimplification, are:

- The vast expansion of economic activity that occurred in the twentieth century and continues today is the predominant (but not sole) cause of the environmental decline that has occurred to date. Yet the world economy, now increasingly integrated and globalized, is poised for unprecedented growth. The engine of this growth is modern capitalism or, better, a variety of capitalisms.
- A mutually reinforcing set of forces associated with today's capitalism combines to yield economic activity inimical to environmental sustainability. This result is partly the consequence of an ongoing political default—a failed politics—that not only perpetuates widespread market failure—all the nonmarket environmental costs that no one is paying—but exacerbates this market failure with deep and environmentally perverse subsidies. The result is that our market economy is operating on wildly wrong market signals, lacks other correcting mechanisms, and is thus out of control environmentally.
- The upshot is that societies now face environmental threats of unprecedented scope and severity, with the possibility of various catastrophes, breakdowns, and collapses looming as distinct possibilities, especially as environmental issues link with social inequities and tensions, resource scarcity, and other issues.

- Today's mainstream environmentalism—aptly characterized as incremental and pragmatic "problem solving"—has proven insufficient to deal with current challenges and is not up to coping with the larger challenges ahead. Yet the approaches of modern-day environmentalism, despite their limitations, remain essential: right now, they are the tools at hand with which to address many very pressing problems.
- The momentum of the current system—fifty-five trillion dollars in output in 2004, growing fast, and headed toward environmental disaster—is so great that only powerful forces will alter the trajectory. Potent measures are needed that address the root causes of today's destructive growth and transform economic activity into something environmentally benign and restorative.

In short, my conclusion, after much searching and considerable reluctance, is that most environmental deterioration is a result of systemic failures of the capitalism that we have today and that long-term solutions must seek transformative change in the key features of this contemporary capitalism. In Part II, I address these basic features of modern capitalism, in each case seeking to identify the transformative changes needed.

The market. In Chapter 4, I focus on the need to transform the market to make it work for the environment, reversing the historical pattern. I examine the urgent need to take seriously neoclassical environmental economics with its emphasis on achieving environmentally honest prices and correcting other market signals, and look at the need to restrain "market imperialism" and excessive commodification.

Growth. In Chapter 5, I focus on what has been called the "growth fetish" and on taking seriously the field of ecological economics, including its critique of endless economic growth and its concern that advanced industrial economies may have already exceeded their optimal or sustainable scale. I explore the dimensions of a "post-growth society," where neither nature nor community is sacrificed to the priority of economic growth. In Chapter 6, I develop the idea that today's economic growth in affluent societies is not materially improving human happiness and satisfaction with life and is a poor way to generate solutions to pressing social needs and problems. I call for alternative measures that directly address these social challenges, which now desperately need attention.

Consumption. In Chapter 7, I focus on materialism and consumerism in today's affluent societies—what has been called our affluenza—and suggest ways to encourage both green consumption and living more simply.

The corporation. In Chapter 8, I take up the challenge to the dominance and power of the modern corporation, including that offered by what is often referred to as the antiglobalization movement, and set out a program to transform corporate dynamics.

Capitalism's core. Chapter 9 is more speculative. Is there something beyond both capitalism and socialism? If so, what might be the dimensions of a nonsocialist system beyond today's capitalism?

In Part III, I consider two potential drivers of transformative change:

A new consciousness. In Chapter 10, I focus on the prospect for profound change in social values, culture, and worldviews. I explore how today's dominant values contribute abundantly to social and environmental alienation and what might lead to a new consciousness that gives priority to nonmaterialistic lives and to our relationships with one another and the natural world.

A new politics. In Chapter 11, I address the search for a new and vital democratic politics—one premised on addressing America's growing political inequality and capable of embracing neglected environmental and social needs and sustaining the difficult actions needed. I examine the vital longer-term goal of strong democracy as well as the immediate steps needed to forge a new environmental politics. An important question in this regard is whether a popular movement that can drive real change is being born.

Taken together, the proposals presented in the chapters that follow would, if implemented, take us beyond capitalism as we know it today.